Franciane Schio
Mariana P. L. C. Carvalho

Perceptions of log transport drivers in the Legal Amazon

Franciane Schio
Mariana P. L. C. Carvalho

Perceptions of log transport drivers in the Legal Amazon

Study of the working conditions of drivers transporting raw timber in the north of the state of Mato Grosso-Brazil

ScienciaScripts

Imprint
Any brand names and product names mentioned in this book are subject to trademark, brand or patent protection and are trademarks or registered trademarks of their respective holders. The use of brand names, product names, common names, trade names, product descriptions etc. even without a particular marking in this work is in no way to be construed to mean that such names may be regarded as unrestricted in respect of trademark and brand protection legislation and could thus be used by anyone.

Cover image: www.ingimage.com

This book is a translation from the original published under ISBN 978-3-330-76707-2.

Publisher:
Sciencia Scripts
is a trademark of
Dodo Books Indian Ocean Ltd. and OmniScriptum S.R.L publishing group

120 High Road, East Finchley, London, N2 9ED, United Kingdom
Str. Armeneasca 28/1, office 1, Chisinau MD-2012, Republic of Moldova, Europe
Managing Directors: Ieva Konstantinova, Victoria Ursu
info@omniscriptum.com

Printed at: see last page
ISBN: 978-620-8-55114-8

To my parents, Luiz and Dilene, who, despite my failings, never stopped believing in me... And when everything showed the opposite, they told me that I was capable...

To my grandparents and godparents Gema and Laudyr for being a reference point in my life, examples of uprightness and honesty. For the affection and love they have for me.

DEDICATE

CONTENTS

SUMMARY

The aim of this book is to understand the real working environment of forestry vehicle drivers, covering topics such as the use of safety equipment, ergonomics, hours worked and the state of repair of the vehicle fleet. It provides relevant data for professionals in the forestry, health and occupational safety sectors, as well as looking at job satisfaction, driver productivity and vehicle conditions. The research was carried out in the municipality of Sinop in the Brazilian state of Mato Grosso, but covered the whole of the north of the state because the municipality is a centre for receiving raw materials, so the interviewees were from surrounding municipalities. The work includes a photographic collection of the vehicle fleet to help visualise and classify the vehicles. The results showed that the drivers interviewed had a low level of education. Most said they only wore their seatbelts near motorway checkpoints. The average body mass index showed excess weight. The average number of hours worked per day was high and the average age of the fleet was older.

Keywords: driver profile, forestry transport, drivers, safety, ergonomics, body mass index

CHAPTER 1

INTRODUCTION

Transport is defined as an intermediary service that moves cargo between different locations, contributing to the development and sustainability of the economic system. Forestry transport is defined as the movement of wood and its derivatives from the forest or factory to the consumer centre (MACHADO, 2009). Or the forestry transport operation consists of moving timber from the forestry yards or roadsides to the place of consumption (MACHADO et al., 2000). Taboada (2002) says that transport is the most important cost element for the vast majority of transport companies, as freight usually accounts for around 60% of logistics costs.

The forestry sector has been one of the most important in the Brazilian economy. In 2003, it contributed 4% of GDP, 8% of national exports, collected US$ 2 billion in taxes and created 2 million jobs (SBS, 2004). And this sector has great growth potential. This is demonstrated by the investments made by forestry-based companies, both in the industrial area, expanding their capacities, and in the purchase of new areas for forest plantations and the acquisition of more efficient machinery and equipment, always seeking to optimise the production process, from planting and maintenance to harvesting and forest transport. All this with the aim of minimising costs through economies of scale (NOCE et al., 2005).

The forestry sector is therefore in full growth and expansion mode, as the demand for products has increased considerably in recent years. This shows the importance of the quality of forestry transport, since it is responsible for a large part of the costs along the yard-to-factory production chain. In forestry, harvesting and transport are the most important stages from an economic point of view, where the final cost of timber can account for more than 50% of costs (MACHADO, 2008).

ICV - Instituto Centro de Vida (2006) says that for the state of Mato Grosso, one of the key sectors for development is the forestry sector. The Portal da Amazônia region has a total of 1,592 companies that process 8 million m^3 of roundwood, resulting in 3.5 million m^3 of processed product, with gross revenue of US$ 673.9 million. In 2004, the timber industry generated approximately 109,000 indirect jobs, representing 6% of the state's economically active population.

Like other sectors of the Brazilian economy, the forestry sector varies according to the fluctuations of the domestic and foreign markets, so anything related to increasing production and lowering costs for producers is of the utmost importance to keep the country competitive and within labour standards.

As far as forestry potential is concerned, Brazil has plenty of land and climatic conditions that, combined with technology and research, can enable the country to stand out in the market. However, we still need a lot of improvement in various areas, from investment, training, studies, processing and transport.

The forestry sector is heavily involved in national agribusiness, whose performance drives the Brazilian trade balance, accounting for almost all (91%) of the surplus seen in recent years. It should be noted that wood transport is closely linked to the production of pulp and paper, solid wood and its derivatives, with a share of approximately US$ 4.24 billion. This figure corresponds to approximately 5.8 per cent of total Brazilian exports and 15.8 per cent of agribusiness exports (BRACELPA, 2011).

Investing in adapting the various stages of the production chain is interesting because it stimulates employee performance, thus achieving greater quality and efficiency in production and consequently generating greater profits for the company. Over time, the tendency is to focus even more on environmental risk prevention programmes, safety and quality of life in forestry work, due to the growing demand from countries to import wood of known origin, i.e. certified wood.

Forestry production in the northern region of the state of Mato Grosso is transported almost entirely by road, and in this context the ergonomic profile of the driver and analyses of the conditions of his machine are addressed in terms of quality of life and performance at work.

Therefore, in view of the content presented, the research problem is the quality of the driver's working conditions and a survey of data relating to machinery.

It is understood that the northern region of Mato Grosso has great economic potential in the timber sector, requiring the forestry transport process to be carried out with maximum efficiency and minimisation of costs, so as not to make final production more expensive.

Based on this situation, it was found that certain procedures should be carried out in relation to vehicles and their maintenance, and the issue of workers' quality of life should also be analysed, based on the environmental risk prevention programme.

There is a need to find out more about the profile of road forestry activity in the municipality of Sinop-MT, as it is directly linked to accidents, incidents, vehicle problems and infrastructure.

As demand for forest products increases, working conditions in the transport sector must also be improved. If various preventative measures are taken in relation to the vehicle, logistics costs can be reduced and thus become less expensive for the wood products industry.

The proposal is therefore justified by the need to find out more about the profile of forestry road transport operators, thus providing a means of identifying needs and making suggestions for the problems that may be diagnosed.

CHAPTER 2

LITERATURE REVIEW

2.1. Forestry transport

As Machado (2009) describes, means of transport are indispensable because they reduce travelling time and allow products to be exchanged between the most diverse communities. A precarious transport system becomes one of the biggest obstacles to society's socio-economic progress. Transport is the activity that gives utility of place, i.e. the right resource in the right place, and creates utility of time, being the right resource at the right time as it reduces the transit time of these resources. What's more, transport establishes the geographical extent that can be reached, known as the radius of action. Therefore, resources are only useful if they are in the right place at the right time, regardless of their distance.

The forestry sector is highly dependent on transport, which can increase final costs. It is essential to have good quality logistics in order to meet deadlines and replenish raw materials in the timber industry.

It is a very important stage for this activity to be carried out, both from the forest to the industry and from the industry to the end consumer. Transport is an intermediary consumer service that provides the movement of cargo between different locations, contributing to the development and sustainability of the socio-economic system (MACHADO, 2000, apud Melo, 2009).

According to Machado et al. (2000), three forestry transport methods prevail: waterway, road and rail. Due to technological developments, pipeline and air transport methods have emerged, but because they are not economically viable, they are rarely used. Some companies transport part of their timber by rail and by river transport, which is more restricted to the Amazon region (SEIXAS, 2001).

Machado (1984, apud, Melo, 2009) also comments that forestry transport is mainly carried out by lorry. The situation has changed little in recent times, with road transport remaining predominant.

Transport integrates timber producers with processing, offering raw materials to forestry industries, and when the market heats up, requiring an increase in the number of timber suppliers, the regional market is stimulated.

Veríssimo et al. (2002) describes the procedure in the logging area, where tracked tractors are used to open up a large network of roads and yards. Then chainsaw operators fell the trees of economic interest. The tractors then drag the felled logs to the yard, where they are separated into logs between 6m and 8m long. Finally, the logs are lifted mechanically and placed on logging lorries to be taken to the region's sawmills.

1.1.1. Types of forestry transport

According to Quadros (2004), transport has always been the necessary impetus for development. This can be seen in the evolution of air, rail and sea transport that exists today. What is truly impressive is the rapid progress of evolution and the achievements that have followed in recent years. We started with a simple small lorry with a load capacity of 5 tonnes and have now reached the technology of lorries with a total weight of up to 74 tonnes, with highly economical and efficient engines.

Waterway transport (fluvial) according to Quadros (2004) is carried out on rivers or groups of rivers, whose depth and width allow for the traffic of vessels such as ferries, boats, ships and even rafts made up of the logs themselves. It is transport used at sea, as it allows the use of larger vessels. The transport of timber by waterway, in the form of logs, is widely used, especially in the Amazon region, where there are a large number of rivers that are large enough and deep enough for this activity.

Inland waterway transport (Figure 1) is competitive in economic terms compared to rail and road transport, as it requires low initial investment in road preparation, offering high load capacity in relation to energy consumption and equipment durability, but is restricted by the existence of navigable rivers and canals (MACHADO, 2000).

When transport by natural means is not possible, there is also an artificial way, which is a navigation channel built by dredging rivers or building dams and dykes. This type of channel is generally used to transport wood chips and, occasionally, logs. This type of transport is substantially cheaper than other methods, but only justifies its use when a large volume is expected to be transported, as it is an inflexible method (QUADROS, 2004).

Waterways have definitely entered the history of transport in Brazil. The completion of the Tietê-Paraná waterway and other waterway projects in the north, southeast and south allow us to envisage a scenario in which agricultural and agro-industrialised products can opt for ports with lower costs and get their products to national consumer centres at lower prices, or even to those centres that were not previously supplied (MARTINS, 1999).

Figure 1 - River transport of timber.
SOURCE: IBAMA (2012).

Rail transport (Figure 2) is used for long distances and large quantities of cargo, generally of low commercial value. This type of transport is interesting because it has competitive costs, since the railway is able to provide these conditions, otherwise the business would not be economically viable. Normally these roads are built within very strict limits, the percentage of ramp has to be close to 1%, because the train can't stop or it skids uphill, especially on rainy days (QUADROS, 2004).

According to Machado (2000), one of the disadvantages of railways in Brazil is the diversity of gauges, the small relative length of the rail network and its consequent inflexibility, restricting its operation to a few large industrial and commercial centres.

Few companies used to use rail transport in Brazil, but today 65% of freight is transported by road. The forestry sector depends even more on this means of transport, taking advantage of the system of paved roads that connect all regions of the country (MACHADO, 2000). Rail transport would be more viable due to its many advantages, but mainly due to its low cost compared to other means of transport, but unfortunately Brazil has a poor infrastructure for this type of product transport.

Figure 2 - Railway trains used to transport timber.
SOURCE: JABLONSKY et al. (1940).

In cable car transport, the load is transported continuously between two static points. A rail is suspended from a network of poles distributed along the route that takes the transported cargo to the industry, or to a yard, where it is then loaded and transported to the processing industry by another type of transport. This system is driven by a closed circuit in which the trolleys are constantly fractionated by means of a steel cable (QUADROS, 2004).

Using air transport (Figure 3) in regions or management systems where traffic is not suitable for building roads, railways or pipelines can be a way out (QUADROS, 2004). Machado (2000, apud Melo, 2009) mentions that this type of alternative model uses balloons and helicopters, but this transport is used in developed countries. This transport model is also suggested when working in an area with a very rugged, mountainous terrain where it is difficult for lorries to enter.

The main advantages of using helicopters are: little dependence on the volume of wood per area; irrelevant damage caused to the wood; minimisation of drag lines;

adaptation to almost any location (O'LEARY, 1962; SEIXAS, 2002); possibility of moving loads vertically; precision in placing the load or accessories; rapid operational cycles; possibility of moving with a load facing winds of up to 90km/h; ease of landing or waiting in the event of a technical, operational or climatic emergency (GUIMIER; WILLBURN, 1984; SEIXAS, 2002; ZAGONEL, 2005); ease of overcoming large obstacles such as mountains, steep valleys, rivers, swamps and sandy areas; and low environmental damage compared to other methods of timber extraction and transport.

The use of this method is only plausible if the product of the exploitation pays for its high cost, for example, sustainable management of huge native forests of valuable species (QUADROS, 2004).

Figure 3 - Aerial transport of timber by cable.

SOURCE: FOREST TRANSPORT AND HARVESTING LABORATORY.

2.1.2. Forestry road transport

Road transport (Figure 4) is the most widely used mode of freight transport in Brazil, helping significantly to shape the costs of various segments of the economy (VELLOSO et al., 1997). Cargo transport in Brazil is highly dependent on road transport. This overdependence is evident when you look at the share of road transport in other countries of continental dimensions. While in Brazil road transport accounts for around 60% of the freight transport matrix, in the United States the share of roads is 26%, in Australia 24% and in China only 8% (Centro de Estudos em

Logística - CEL et al., 2002).

The importance of lorries as a means of transport is due not only to the volume of cargo to be transported, but also to their versatility or ease of movement and interconnection between points of origin and destination, located almost everywhere on the earth.

As you can see, forestry transport is the movement of timber from the logging yards and roadsides to the processing plantations. In Brazil, it can be carried out by various means (rail, pipeline and road), with the latter accounting for 85% of all timber transported and 62% of all products transported in the country. Even with the infrastructure problems faced by most roads, it is the only means of connecting industries to their wood supply sources, covering various areas within Brazil (STEINet aL, 2001).

As we have seen, timber transport in Brazil is carried out by road, and is predominantly responsible for the largest share of timber put into mills. It is a sector that is often under pressure to increase costs due to the installation of toll stations on the highways, stricter inspection in relation to the "Scale Law", and the sector is also fundamentally affected by the readjustment of fuel prices (SEIXAS, 1992). Today this law has been reformulated and replaced by more up-to-date legislation, but the aim is the same: to set limits on weights and dimensions.

In our country, road freight transport earns more than R$40 billion a year and moves 62% of all freight. However, this type of transport stands out due to the large number of ways of working, as it operates with more than 350,000 autonomous hauliers, 12,000 transport companies and 50,000 own-load hauliers. The main factor for this is the ease with which competitors can enter the sector, due to the lack of regulation, which ends up increasing the range of services on offer, reducing freight costs. However, the most rational way to reduce this cost is to increase the use of the fleet, without jeopardising the level of service (MACHADO, 2009).

The transport of wood to the industries is currently basically done by road (SIMÕES, 1981). While in the 1980s the author made this claim, today this reality has not changed. Few companies used rail transport in Brazil, and today 65% of freight is transported by road. The forestry sector depends even more on this means of transport, taking advantage of the system of paved roads that connect all regions of the country (MACHADO, 2000, apud Melo, 2009).

It is the means of transport that is of most interest to the Brazilian forestry sector because of the political decisions made in this sector in the past. The system currently in use in Brazil is made up of different types of vehicles with different load capacities (QUADROS, 2004).

Machado (2000) highlights the great advantage that road haulage vehicles have over other means of transport: the possibility of moving goods "yard to yard", as well as the choice of different routes and load capacities on offer.

He also points out that the main means of forestry transport is by lorry, with road transport taking precedence due to the importance that these vehicles have acquired and that they are considered to be one of the most significant phenomena today.

This importance stems not only from the high volume of cargo moved between the producing, intermediary and consuming companies, but also from the fact that it is indispensable in connecting points of origin or destination of goods (MACHADO, 1985, apud Melo, 2009). The fact is that, with the technological evolution of heavy vehicles, there are new options ranging from simpler compositions, such as Romeo and Juliet (lorry with one trailer) to the truck (lorry with two trailers) or road train (mechanical horse plus two semi-trailers or one semi-trailer and one trailer) (SAAB SCANIA, 1985, apud Melo, 2009).

The transport of logs, which is very common in the forestry sector, must also be carried out in accordance with CONTRAN regulations (CONTRAN, 2006, apud Valeriano 2009). Resolution No. 246 of 25 July 2006 sets out the safety techniques required for the transport of raw wood logs by road freight vehicles. In this case, the logs must be transported in the longitudinal direction of the vehicle, in a pyramidal (triangular) arrangement. Whichever way the logs are arranged, the vehicles must have, among other items, suitable stanchions, steel cables or tensioned polyester straps with a ratchet system.

The Brazilian road fleet has an average age of 17.5 years and locomotives with an average age of 25 years; roads in very poor, bad or deficient condition in 78% of cases; low availability of rail infrastructure and waterways that are still little used

(BIODIESELBR, 2006). With regard to length, based on the availability indicator, measured by the ratio "total kilometre of road per territorial space (in km^2)", the supply of transport routes in Brazil is equivalent to 69% of that in China, 55% in Canada, 45% in Mexico and 6% in the USA (CEL et al., 2002).

Quadros (2004) shows that lorries are classified according to their size, load capacity and composition. According to articles 99 and 100 of Law No. 9.503 of 23 September 1997 (Brazil, 1997), no vehicle or combination of vehicles may be driven in excess of the passenger capacity, Total Gross Weight (TGW) or Total Combined Gross Weight (TCCW), or with an axle weight above that set by the manufacturer, and must not exceed the Maximum Traction Capacity of the tractor unit.

Figure 4 - Vehicle used to transport timber by road.

SOURCE: Photos by the author (2012).

2.1.3. Cargo vehicle combinations

Coelho (2010) gives a presentation of the different types of lorries, their specifications and capacities, starting with small urban cargo vehicles, up to the large articulated lorries that are seen on highways, used to transport large quantities of cargo over long distances. CONTRAN (the National Traffic Council) limits the maximum weight per axle that vehicles can carry. This limit is due to the fact that the greater the force the tyres apply to the asphalt layer, the greater the degradation of this asphalt. Therefore, lorries can carry a lot of weight, as long as it is well distributed over several axles (more wheels to distribute the weight of the load). Machado (2000) categorises vehicle types:

- Truck: consists of a single tractor unit (mechanical horse) with 4x2, 4x4, 6x2 or 6x4 traction.
- Articulated (trailer): made up of a tractor unit (mechanical horse) with 4x2 or 6x4 traction and a semi-trailer.
- Conjugate (bimini): formed between a lorry and a trailer (lorry+trailer).
- Bitrem: combination of a mechanical horse and two semi-trailers (mechanical horse+semi-trailer+semi-trailer).
- Tritrem: combination of a mechanical horse and three semi-trailers (mechanical horse+semi-trailer+semi-trailer+semi-trailer).
- Rodotrem: combination of an articulated vehicle and a trailer (mechanical horse+semi-trailer+trailer).
- Truck: combination of a lorry and two trailers (lorry+trailer+trailer). The vehicle combinations are illustrated in Figure 5.

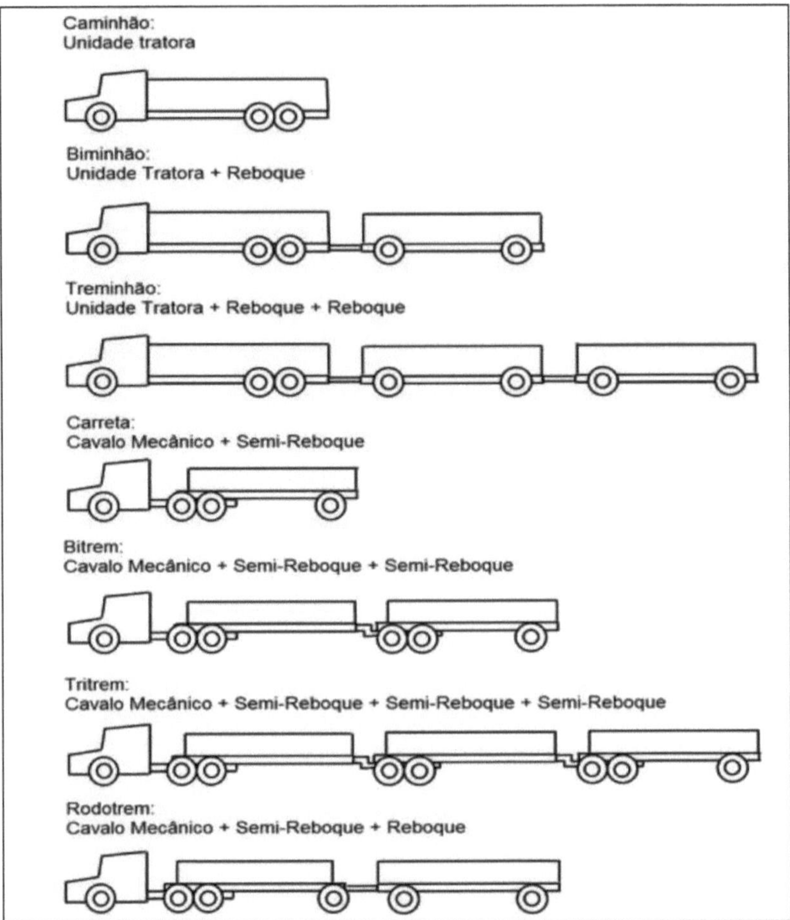

Figure 5 - Schematic of conjugated vehicle combinations.
SOURCE: MACHADO (2000).

2.2. Forestry transport in the Legal Amazon region

Roads play an important role in a country's economic and social development. In Brazil, road transport is considered to be the main route for integration and is essential for transporting industrial and agricultural production (TRINDADE et al, 2005). The Brazilian road network is around 1,725,000km long, with approximately 10% paved; of this total, 83% fall into the category of municipal and vicinal roads, of which 1.2% are paved (VELTEN et aL, 2006).

Due to its characteristics in terms of load specification and freight exclusivity, forestry transport allows lorries to travel loaded in only one direction, which means that costs tend to be higher per unit of volume than in other types of transport (SEIXAS, 1992).

Forestry transport consists of moving wood from the yards or roadsides in the stands to the place of consumption or the company yard. Even with the precarious traffic conditions found on most roads, these are the only means of connecting industries to their raw material supply sources, covering distant areas within Brazil. For this reason, it is important to carry out studies to improve the efficiency of forestry road transport (SILVA et. al., 2008).

According to CONTRAN Resolution 188 of August 2006, among other provisions, the transport of loads of logs must comply with various safety requirements. Logs must be transported in an orderly fashion, arranged longitudinally or transversely on the vehicle body. The height of the load must not exceed that of the panels, stanchions and side guards of the vehicle body. It must contain: front and rear panels of the vehicle body; metal side struts, perpendicular to the plane of the floor of the vehicle body (stanchions), with at least two stanchions required for each log or bundle of logs; steel cables or polyester strapping, with a minimum tensile breaking capacity of 3000kgf, correctly tensioned by ratchets or a self-adjusting pneumatic system, fixed to the body.

However, it is notable that in the north of the state of Mato Grosso and the south of Pará there is a lack of mandatory vehicle safety equipment, especially in older vehicles, and the further into the interior of the country you go, the worse the situation gets. Figure 6 shows a vehicle with adequate protective equipment.

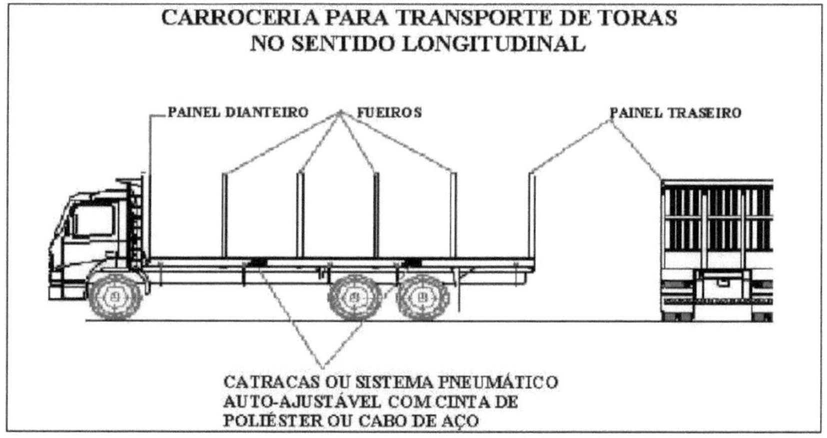

Figure 6 - Safety equipment for forestry transport.

SOURCE: CONTRAM (2006).

Magri et al. (2005) state that with the opening of other roads in the 1970s and 1980s, Mato Grosso and the Amazon in general began to integrate into the national economy. The biggest development was in Mato Grosso, which became a major agricultural production centre in the last decade of the 20th century. The region has also gone through various types of economic cycles, including mining, timber and agriculture, which can be subdivided into the livestock and agriculture cycles.

Logging was the first activity in the region and continues to feature prominently in the regional economy, whether in the Amazon rainforest or transitional areas. In many places, such as Sinop, sawmills continued to operate in the region even after the arrival of agriculture. However, with the depletion of raw materials and the start of stricter inspections to ensure that the industries obey the law in force, many of them closed down or moved further north to Mato Grosso or south to Pará.

2.3. Drivers of forestry cargo transport vehicles

In Brazil, the internal transport of agricultural products, industrialised products, raw materials, passengers and others is carried out almost entirely by road , resulting in a large number of trucks and buses (KILESSE, 2006). These drivers differ in certain respects, because when transporting grain by road, i.e. when transporting agricultural cargo, the lorry goes back and forth loaded and covers long distances between trips, as is the case, for example, when transporting grain from the north of MT to ports in the south of the country. Another difference that has also been noted is that timber cannot be transported all year round; it is harvested seasonally.

Another distinguishing feature of forestry is that there are many old vehicles on the road, even though this practice goes against the advice of authors specialising in the area. For example, one author states that lorries should not be older than 10 years (MACHADO, 2000).

Forestry vehicles need more power due to the use of larger vehicles to make transport viable. In 2005, 40 per cent of the vehicles used had power between 360 and 380 hp (PEREIRA, 2005).

Another concern for drivers is fuel economy, which is directly related to the energy efficiency of each vehicle, and is set to increase with load capacity (MACHADO, 2000).

According to Pereira (2005), the transport distance between the raw materials in the stockyard and the industry in 2005 was an average of 100 kilometres. It is therefore clear that when these distances increase, it is due to a shortage of raw materials.

Outsourcing has become a current trend for companies of all sizes and segments, in order to remain competitive in the market (MACHADO, 2000).

Valeriano (2009) explains in his Primer on forestry work that this sector has its own characteristics, different from other sectors. Working conditions and environments vary and have a major impact on labour issues and workers' well-being.

Some of the questions used to understand the profile of these workers are cited by Minette (1996), who shows that some of the main questions used to characterise the profile of operators are: age, weight, height, body mass index, working hours, marital status, schooling and work shift.

2.3.1. Analysing the profile of forestry vehicle drivers

It is very important for ergonomics, health and safety to know the profile of workers, which is necessary to make changes to working conditions in order to achieve health, safety, comfort, efficiency and quality of life (SILVA, 2007). The origin of the term "ergonomics" comes from the Greek words "ergos" (work) and ***nomos*** (natural laws) (MURREL, 1975). Thus, ergonomics seeks to adapt work to man within a humanistic approach, as opposed to the mechanistic approach that aims to adapt man to work. Expanding this definition a little further, we could say that "it is easier to adapt the environment to man than man to the environment" (ENGLAND, 1981).

The term quality of life implies the more or less harmonious interrelationship of the various factors that mould and differentiate the daily lives of human beings and result in a network of phenomena, people and situations. Many factors of a biological, psychological and socio-cultural nature, such as physical health, mental health, longevity, job satisfaction, family relationships, disposition, productivity, dignity and even spirituality are associated with the term quality of life. Modern society has made life increasingly stressful. In this context, workers' quality of life and health become compromised, which is why it is necessary to achieve a level of excellence in terms of productivity and the quality of the work provided (ALVAREZ, 2012).

The driver profile focuses on variables that influence their working conditions, productivity and satisfaction.

According to the IEA (International Ergonomics Association), ergonomics applied to work (or human factors) is a scientific discipline concerned with understanding the interactions between human beings and other systems and applying theories, principles, data and methods to projects in order to improve the quality of human work life and the overall performance of the system (ABERGO, 2006, apud Machado, 2008).

Ergonomics is an applied science that encompasses eclectic and subsidised concepts from social, human and exact sciences, as well as technology, to adjust work to the physical and mental conditions of human beings, in order to create good conditions for their satisfaction, health, safety, productivity at work and well-being.

In the man-machine system, one of the aims of ergonomics is to adapt work to the human being. In order to carry out any task more efficiently, man generally resorts to machines or auxiliary tools (MACHADO, 2008).

According to Machado (2008), the performance of a man-machine system depends on the individual characteristics of the operator, such as their anthropometric measurements, age, training and motivation; the environmental conditions, such as lighting, climate, noise and vibration; and the condition of the machine, i.e. size, power and state of repair.

For Fernandes (1996, apud, Machado, 2008), quality of life at work in the world of work is called the variable quality of work life - QWL, a term that gained acceptance in industry and then moved on to other productive sectors. It says that there are factors that affect people's quality of life when they are working and that these will provide favourable conditions that are essential for better performance and productivity.

According to Silva (2007), the national forestry sector expanded greatly from the 1990s onwards, earning it a prominent place in Brazil's economy. However, the success of the economic indicators has not been accompanied by improvements in living, health and working conditions. The northern region of the state has an old, poorly equipped fleet, which, combined with roads that are not properly maintained, produces a lot of shaking, and the old lorries do not offer the thermal and acoustic comfort that the new ones do.

Changes in forestry transport must be based on ergonomic principles in order to offer better working and health conditions for workers, thus promoting comfort, well-being, safety and productivity in the workplace.

When individuals are subjected to adopting inappropriate postures or handling poorly designed equipment, taking ergonomic aspects into account, the impact of the force exerted, even for a short period of time and on just one region of the body, can cause severe damage to the worker's muscular and skeletal systems (FIEDLER, 1995; and CHAFFIN and ANDERSON, 1990).

According to Porto (2000), the idea of worrying about the problems of illnesses and accidents at work should not be restricted to compliance with safety regulations and the provision of personal protective equipment, but also to training, technology and work organisation.

In addition to social concerns, Minette (1996) pointed out that the human being is the main component that determines productivity, as well as the success or failure of a work system, which reiterates the need to guarantee them a healthy and safe environment.

Poor conditions in the working environment have an impact on the quality of life of the worker, the product manufactured and the service provided (SILVA et al, 2002). According to Machado and Souza (1980) the elimination of accidents is of great importance to man. When they are not fatal, accidents generally result in a wide variety of bodily injuries.

While the lorry cab used to be seen as a stressful and uncomfortable place, with a lot of physical work involved in shifting gears, today's lorries are more comfortable, with more modern standard items (air conditioning, power steering, electric windows and locks, automatic gearbox and seat adjustment) that are already part of the driver's routine, although they still have a low market share due to high costs. On the other hand, basic adjustments to vehicle equipment help make driving easier and more comfortable (MANZATTO, 2012).

2.4. Qualitative and quantitative questionnaire

The methodology used in this study was defined by Santos (2008) as qualitative descriptive research. Quantitative research makes it possible to measure opinions, reactions, sensations, habits and attitudes, using a sample that statistically represents the population.

Faleiro (2011) considers quantitative research to mean that everything can be quantified, which means translating opinions and information into numbers in order to classify and analyse them. It requires the use of statistical techniques such as percentages and averages. Descriptive studies, on the other hand, have well-defined objectives, with structured formal procedures aimed at solving problems or evaluating alternative courses of action. The purposes of descriptive studies are: to describe the characteristics of certain groups; to estimate the proportion of people in a group who are likely to act in a certain way and to make specific predictions about some aspect.

2.5. Perception

Perception according to Ramos et al. (2005) suggests that each individual perceives, reacts and responds differently to their environment. The resulting behaviour is therefore the result of the perceptions (individual and collective) of each person's cognitive processes, judgements, expectations and experiences.

CHAPTER 3

MATERIALS AND METHODS

3.1. Characterisation and location of the study area

The research was carried out in the urban perimeter and farms (Figure 7), where petrol stations, mechanics and timber companies were visited in the city of Sinop, with geographical coordinates S 11° 51' 09" O 55° 30' 32". It has a population of 113,099 inhabitants, a territorial extension of 3,942km^2 and is 488km north of the capital Cuiabá in the state of Mato Grosso.

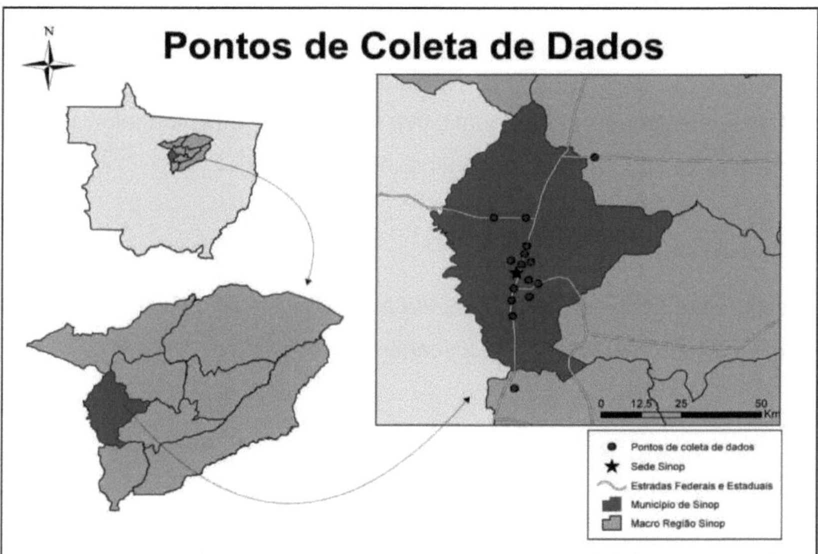

Figure 7 - Location of the study area and data collection points.

3.2. Data collection

The data was obtained through interviews with drivers of vehicles transporting raw timber in logs to supply the region's timber industries.

A qualitative and quantitative questionnaire (ANNEX II) was drawn up and applied randomly to 50 drivers. According to the methodologies of Melo (2009) and Zuanazzi (2011), 30 drivers were interviewed in each survey.

The interviews took place between July and September 2012 and were always carried out in the early hours of the morning, as this was usually the time when the drivers were arriving from the logging area and preparing to unload at the logging yard. These interviews lasted between 1 and 2 hours, as most of the time it was necessary to wait for the cargo to be unloaded.

The questionnaire aimed to include information from the real scenario in order to detect the problems and peculiarities of drivers.

Together with the questionnaire, a photographic archive of the vehicle fleet was drawn up to record the types of vehicle combinations and their characteristics, and the images were recorded using a digital camera.

3.3. Data compilation

From the data obtained in the qualitative-quantitative questionnaire, graphs and tables were drawn up to better understand the results.

3.4. Analysing the data

The results were analysed using simple descriptive statistics, using averages and percentages to help visualise the real scenario.

CHAPTER 4

RESULTS AND DISCUSSION

4.1. Drivers

4.1.1. Socio-demographic information

According to the drivers' personal files, the sociodemographic information regarding the gender, age, level of education and marital status of the 50 interviewees is presented (Table 1).

Table 1 - Socio-demographic information that stood out among the drivers.

Information	Percentages and ranges
Sex	100% male
Age	22 to 70 years old
Level of education	82% primary education
Marital status	62% married

According to the results in Table 1, all the drivers interviewed were male, which was to be expected. Their ages ranged from 22 to 70, from beginners in the profession to men with 48 years' experience on the road. These results were similar to those found by Teixeira (2008), who, in a study of 104 lorry drivers, found that all the drivers were male and aged between 20 and 79.

The majority of drivers had a low level of education, with 82 per cent reporting that they had only completed primary school.

With regard to marital status, 62 per cent of those interviewed were married, while the remainder were 24 per cent single, 10 per cent in a stable union, 2 per cent separated and 2 per cent widowed.

4.1.2. History of professions

When asked about their previous occupations, the drivers reported a wide range of activities (Table 2).

Table 2 - History of professions.

N°	Profession	N°	Profession
7	Machine operator	1	Harvester operator
6	Grain transport driver	1	Boiler operator
5	Loader operator	1	Planer operator
3	T ratorist	1	Sales
3	Farmer	1	Shop assistant
2	Conveyor operator	1	Washer
2	Forklift operator	1	Graphic designer
2	Bricklayer	1	Seed breaker
2	Labourer	1	Sawyer
2	General services assistant in Serraria	1	Delivery man
1	Mechanic	1	Exploration area manager
1	Butcher	1	Paper Industry Loader
1	Boiler Operator	1	Carpenter
1	Sales	1	Agricultural Technician
1	Shop assistant	1	Architecture Projector

According to the history of professions, it was possible to observe that despite the diversity of professions reported, most of the interviewees had worked in some activity related to the timber industry.

4.1.3. Level of satisfaction and permanence in the profession

When asked if they liked their job, of the 50 interviewees, only 1 said they didn't like it. With regard to staying in the profession, the reasons are shown in Figure 8.

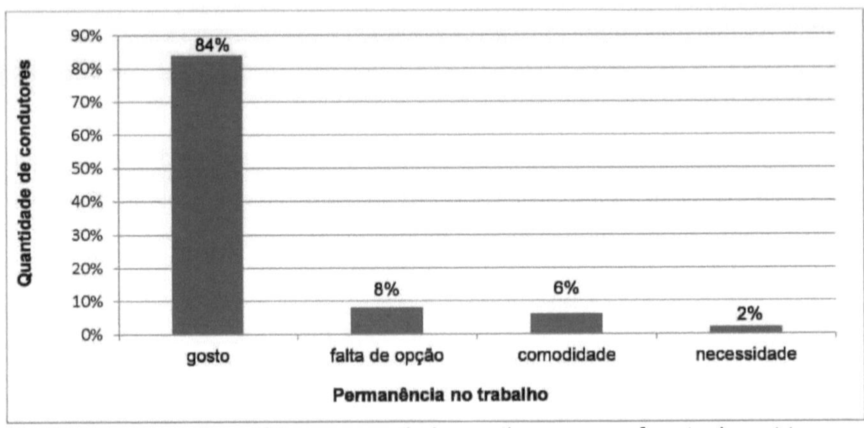

Figure 8 - Distribution of drivers in relation to the reasons for staying at transporting raw timber in the municipality of Sinop-MT.

According to the data in Figure 8, 84 per cent of drivers reported that they were in the profession because they liked it, which is a very representative percentage. It can be said that Brazil made a lot of progress in economic and social terms during the 2000s. This favourable performance was the result of a major advance in social policies, which were able to significantly reduce poverty and inequality (BARROS et al., 2006). In addition to the qualitative improvement in employment as a result of the greater formalisation of employment by 2007. Guaranteeing decent jobs for all workers was a commitment made by the Brazilian government, documented in the National Decent Work Agenda (IPEA, 2007). However, in a country where many inequalities (social, income, access and opportunities) have prevailed throughout history, much remains to be done to ensure that everyone has access to decent work and can remain in this job for as long as possible (GUIMARÃES, 2010).

In terms of length of service, it can be reported that the drivers interviewed had an average of 18 years in general cargo transport, and an average of 14 years as roundwood drivers. Another similar study found an average of 17 years working in general cargo transport and an average of 12 years as a roundwood driver (FELIPPE, 2012).

2.5.1. Training, health, safety and ergonomics

The following are the drivers' perceptions of health, safety, ergonomics and the existence or not of qualification courses for their current profession.

Figura 1 Professional training

Figure 9 shows the training courses taken by the drivers.

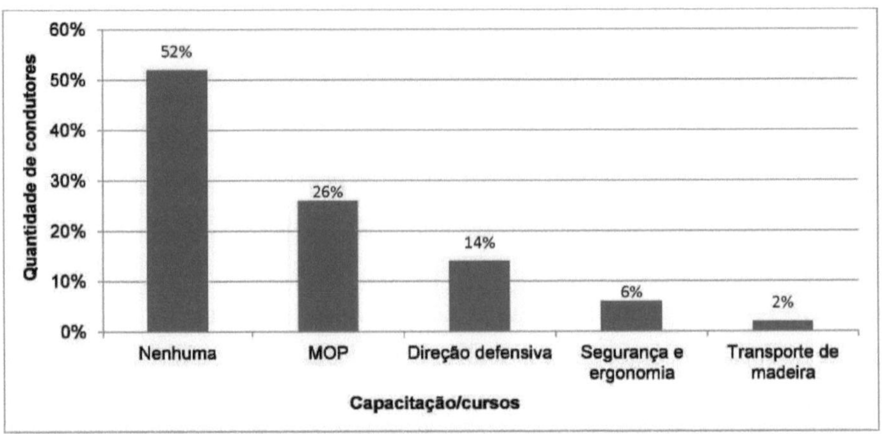

Figure 9 - Distribution of drivers in relation to professional training for transporting raw timber in the municipality of Sinop-MT.

The graph shows that 52 per cent of the interviewees reported that they had never taken any kind of course related to their profession. And 26% had taken the dangerous goods transport course - MOP.

In Brazil, most of the wealth produced is transported from the production centres to the marketing and export centres, predominantly by road. Unfortunately, road accident indicators show alarming figures for freight transport, as shown by a study by the National Transport Confederation (CNT). The study showed that the accident rate in Brazil reached 3.27 accidents/km of road in 1998, around 226% higher than the rate found in the United States. With regard to the rate of road deaths, it is between 10 and 70 times higher than in the seven richest countries in the world (CNT, 2002).

Companies like International Paper have standards for the training of drivers and loaders;

all drivers must be qualified for the job and hold a national driving licence specific to the type of vehicle they will be driving, as determined by the Brazilian Traffic Code, and must also receive Defensive Driving training from a duly qualified professional. All drivers must be retrained annually in Defensive Driving. As well as taking part in the refresher course, it is the safety technician's responsibility to validate this through a written assessment of the course content. All drivers who need to enter the plants must be trained and fully aware of the safety instructions and Emergency Plans at each unit (MARCHIORI, 2009).

Figura 2 Health

Figure 10 shows the data on the drivers' Body Mass Index.

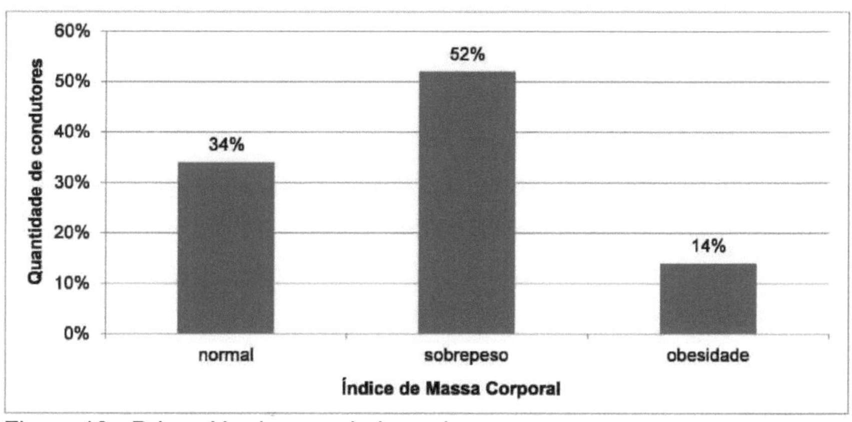

Figure 10 - Drivers' body mass index values.

The graph highlights the Body Mass Index (BMI), which had an average of 26.28 kg/m^2 indicating overweight among the drivers, a fact that can cause a series of consequences that compromise their health and quality of life. Mcardle et al. (2002) explain that the body mass index (BMI) is a formula that indicates whether an adult is overweight, obese or below the ideal weight considered healthy. The formula for calculating body mass index is: BMI = weight / (height)$^{(2)}$ (*Quetelet*'s formula). This formula is extremely simple, easy to use and safe in terms of scientific standards. This indicator is considered normal between: 18.5 and 25.0kg/m^2, overweight: 25 to 29.9kg/m^2, obese: 30 to 39.9kg/m^2, morbidly obese: > 40kg/m^2.

The graph shows that 34% of the drivers had a normal BMI, 52% were overweight and 14% had grade I obesity. This indicates that measures need to be taken to avoid the next stage of the disease. According to the study by Teixeira (2008), the body mass index classification showed that 28 participants (27%) had a normal BMI, 42 participants (40%) were overweight, 27 participants (26%) were mildly obese, 6 participants (6%) were moderately obese and 1 participant (1%) was severely obese. This study of lorry drivers aged between 20 and 79 also found that the majority were overweight.

In recent decades there has been a rapid and growing increase in the number of obese people, which has made obesity a public health problem. This disease has been categorised as a disorder of high energy intake. However, evidence suggests that much of obesity is due to low energy expenditure rather than high food consumption, while the physical inactivity of modern life seems to be the biggest etiological factor in the growth of this disease in industrialised societies (BATISTA; SILVA, 2005).

Obesity can lead to cardiovascular problems and high blood pressure, and body weight can overload the spine and lower limbs, leading to long-term degeneration such as arthrosis (CZEPIELEWSKI 2001).

Table 3 shows the anthropometric characteristics of the drivers sampled.

Table 3 - Anthropometric characteristics of the drivers.

Features	Average	Minimum	Maximum
Width (m)	1,75	1,60	1,99
Height (m)	1,72	1,55	1,98
Weight (kg)	78,06	50	110
BMI (kg/m^2)	26,29	19,72	31

Anthropometry is defined as the study of measurements of the characteristics of the human body and mainly covers the study of linear dimensions, diameters, weights and centres of gravity of the human body and its parts. Few anthropometric studies have been carried out in Brazil, with some being carried out by research institutions in a non-systematic way (IIDA, 2003).

The height and wingspan of the drivers were measured. Wingspan is the total length of the open arms. The average height of the 50 workers shown in Table 3 was 1.72 metres and the average wingspan was 1.75 metres. In his research, Correia (2009) states that wingspan is a characteristic associated with height and that measuring wingspan can indicate back problems. The average values were considered normal as they showed little difference, with a 3cm increase in wingspan.

Figure 11 shows drivers who used to drive at night.

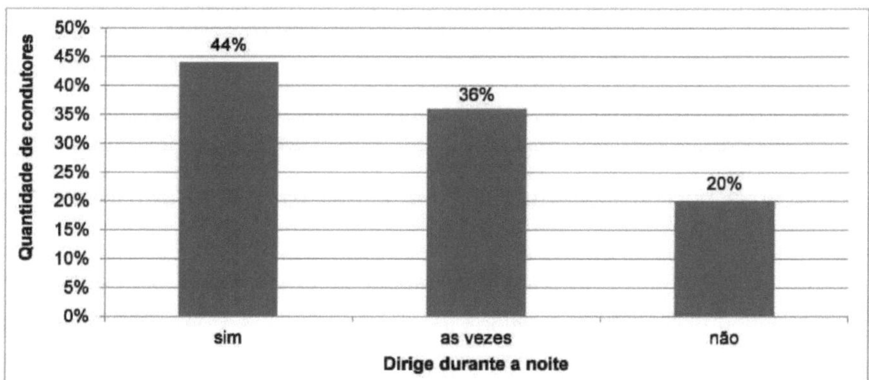

Figure 11 - Distribution of drivers who drove at night.

As we discussed, we found that 44% of the drivers drove at night and 36% sometimes, in other words, in general, 80% of the drivers drove routinely or not at night. This habit occurs for a number of reasons: many drivers spend the night in the logging area and then return to the logging company at dawn. This preference is because there is less traffic on the roads, and some reported that they do this to avoid inspection, for various reasons, such as the lack of a note and the excess weight of the load. According to the drivers, there is a petrol station on the MT 220 between the towns of Sinop and Juara, where many drivers stay overnight.

The average number of hours driven per day was around 10 hours, and they preferred to drive in the early hours of the morning. The average time they drove without interruption was around 3 hours, but the vast majority said that they had short breaks to tie the cables that hold the logs, where according to reports the maximum time between one cable adjustment and another is 2 hours.

Through a time study involving loading, unloading, travelling and waiting at the terminals, it was possible to diagnose the total cycle time of a journey, which was around 12 hours (RIVEIRA, 2012).

Freight transport drivers are a professional category that still faces situations where they have to work up to 20 hours straight or more. The working hours of these professionals, in the words of Paulo Douglas Almeida de Moraes, the Public Prosecutor for Labour in the state of Mato Grosso, are "inhumane, barbaric and cruel", due above all to the intensification of work, which is also caused by the prior determination of delivery times and dates for the goods being transported (AMARAL, 2010).

In the drivers' own accounts, they say that because most of the lorries are overweight, most of them prefer to drive at night or in the early hours of the morning in order to avoid the police. The drivers communicate via amateur radio when they come across an inspection and so they pull over to wait for night or dawn to continue their journey.

Convention 153 of the International Labour Organisation (ILO), adopted in 1979, is legislation on working conditions in road transport. According to it, drivers must have a break every four hours while driving. The working day must not exceed nine hours a day and 48 hours a week (PETZHOLD, 2011).

One point noted by the sample was the excessive working hours, up to eighteen hours a day. The participants justify this by saying that they earn according to the amount of freight they carry out - the more journeys, the higher the monthly income (FELIPPE, 2012).

Petzhold (2011) suggests some procedures for improving working conditions in transport, such as time-controlled journeys, because nowadays, with the benefits of information technology, it would be very easy to allocate a pre-established time for a particular journey, especially at toll stations. Interstate bus companies already control the maximum journey time for their drivers. So do some lorry companies. The problem persists with autonomous drivers, who don't respect these limits. The author also emphasises the need for rest time. Human beings need to have daily breaks (rest periods should not be less than eight hours), weekly and annual rests (holidays). Drivers with formal licences certainly enjoy this right. The problems are autonomous drivers, due to a lack of supervision. With the above items being complied with, fatigue would be controlled and safer driving would be achieved, especially if there was a concern to add a defensive driving course for professional drivers. What we see today among the majority of self-employed drivers is a total lack of concern for these items, to which must be added the abusive use of "arrebites", "sleep suppressants", thus becoming a very serious condition for the problem of fatigue (PETZHOLD, 2011). The approval of a specific Regulatory Standard (NR) for road transport workers, in view of the items presented above, would be beneficial. As well as funding for monographs, theses, research, films, on safety and health at work (particularly in traffic) (PETZHOLD, 2011).

c) Security

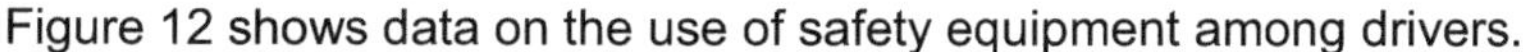

Figure 12 shows data on the use of safety equipment among drivers.

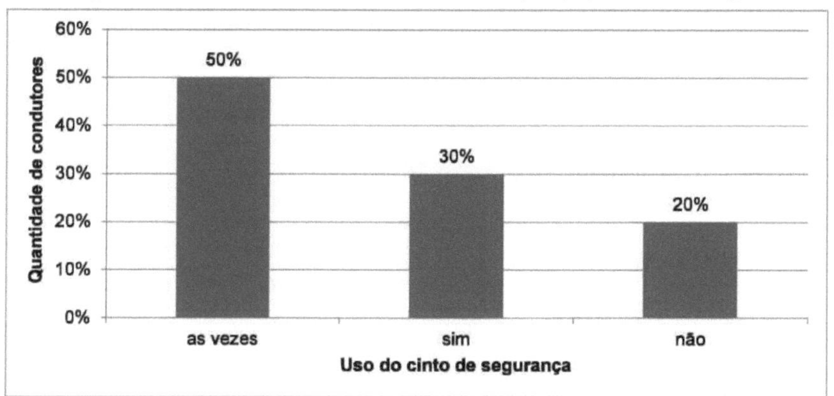

Figure 12 - Distribution of drivers in relation to the use of seat belts when transporting raw timber in the municipality of Sinop-MT.

The figure shows the lack of seatbelt use, as 20% said that they don't wear them and 50% that they wear them sometimes. The majority said that they only wear seatbelts on the federal motorway and near checkpoints.

The seat belt is a device for preventing physical harm to the driver of the vehicle. According to Article 65 of the Brazilian Traffic Code, seat belts are compulsory for drivers and passengers on all roads in the country. The seat belt is a simple tool that serves to protect life and reduce the consequences of traffic accidents. In the event of a collision, it prevents the human body from hitting the steering wheel, dashboard and windscreen, or from being thrown out of the car.

Based on a simulation carried out by the testing department of one of the country's largest car factories, if a 60kg person is thrown from a car at a speed of 60km/h, this body will cause a collision of approximately one tonne, making it impossible for instant death not to occur (CÂNDIDO, 2009).

It is up to the government, BRASIL (2000), to permanently evaluate the National Traffic System, which includes the municipalities exercising their competences described in article 24 of the Brazilian Traffic Code; establishing guidelines for the National Traffic Policy and "monitoring compliance with it"; emphasising the behavioural dimension, in the various roles played by man in the flow of traffic. As "subsidies for the formulation of public policies" (BRASIL, 2003), Table 4 shows the data on traffic accidents that occur during freight transport.

Table 4 - Data on traffic accidents during freight transport.

Traffic accidents	Yes	No
Have you ever had a problem with the load	34%	66%
Have you ever been involved in an accident	14%	86%
Deaths or injuries	4%	96%
Consider dangerous cargo	96%	4%

As can be seen in the table, 96 per cent consider cargo to be dangerous, with 34 per cent having had some kind of cargo-related problem, 14 per cent having already been involved in an accident and four per cent having been killed or injured.

Problems with the load included: the load tipping over during the journey; cables snapping; when unloading, a log that hadn't been fitted properly rolled off the vehicle; when removing the log with a shovel, the log came loose and rolled off, hitting the driver; the vehicle was overloaded and/or had a poorly distributed load when travelling on uneven ground and ended up tipping over.

According to Austroads (1994), human error (neuropsychological, workstation, physical fatigue, among others) is responsible for 67% of accidents, with the road and/or the vehicle being the other factors influencing accidents.

d) Ergonomics

Table 5 describes various ergonomic issues.

Table 5 - Driver perception of vehicle ergonomics.

Questions	Yes	No
Is access to the cabin easy?	49	1
Is the work position comfortable?	46	4
Does the seat fit your body structure?	49	1
Is the cabin a comfortable size?	37	13
Is the design and angle of the seat correct?	44	6
Does the seat have cushioning and height and length adjustments?	44	6

Does the level of vibration/trepidation bother you?	14	36
Does the noise level bother you?	10	40
Is there an air conditioning system?	26	24

For a better understanding of the information, data on drivers' perceptions of ergonomics are shown in percentages in the graph below:

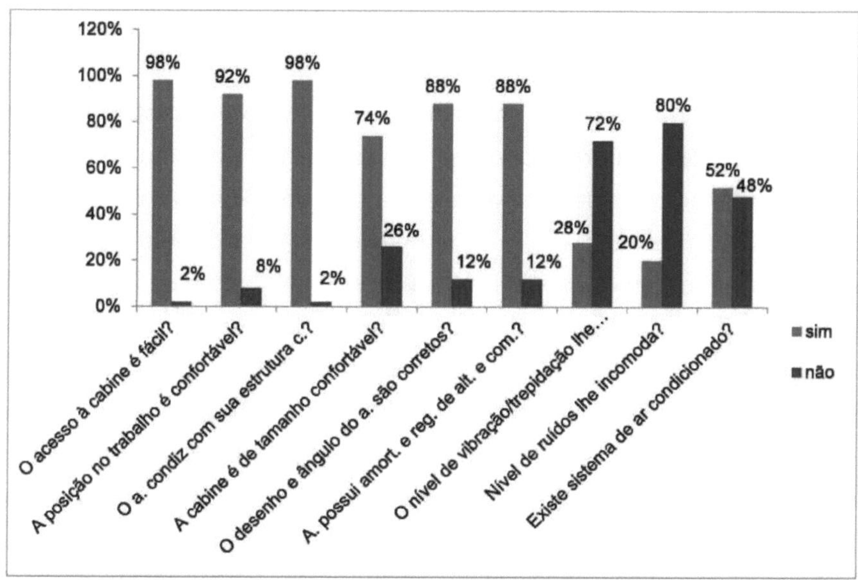

Figure 13 - Driver perception of vehicle ergonomics.

Figure 13 shows the drivers' perception of the safety and ergonomics of the vehicle in which they work. The results were as follows: 98% consider access to the cab to be easy, i.e. the height between the steps is adequate. For 92 per cent, the working position is comfortable. For the 4 drivers who didn't feel comfortable, it is noted that this is due to the vehicle's advanced age, which ranges from 15 to 25 years. The same happened with the question about the seat, when asked if it was in keeping with their body structure, only 1 driver said they were not satisfied with the seat, because their vehicle is from 1977, which indicates that it is outside the parameters of safety and ergonomics.

The design and angle of the seat were correct in 88 per cent of cases. Also in 88 per cent of cases, the seat had cushioning and height and length adjustments.

Regarding seat comfort, Felippe (2012) asked the sample if they were comfortable or not. Around 67.3% reported that the vehicle's seat was comfortable. They reported that the best seat is the pneumatic one, which adapts to the weight of each individual, making it comfortable because it avoids impacts during the journey.

Driving in an uncomfortable position, as well as making the journey unpleasant, can have physical consequences that will still be present at the end of the driver's activity. Even though they come in different makes and models, most vehicles need to follow international standards that mean anyone, regardless of size or weight, can find their ideal driving position (AUTO ESPORTE, 2011).

As for the size of the cab, 74 per cent considered it to be satisfactory. One driver disagreed, adding that the roof of the vehicle was too low. 72 per cent said they were not bothered by the level of vibration and shaking. The level of noise was also not a discomfort for 80 per cent of the drivers. 52% said there was an air conditioning system, of those who didn't at least 5 reported having an inter-climate system.

Given the general age of the fleet, it is admissible that some drivers have many complaints, but the overall result was somewhat positive. It is understood that an ergonomically sound workplace allows for better concentration and driver satisfaction. Seats that fit the body and the control elements help in their responsible task behind the wheel. Also, in the case of high temperatures, a powerful air conditioning system would quickly and permanently ensure a pleasant temperature in the cab. This is where dissatisfaction was greatest, with 48 per cent of the cases stating that there was no cooling system, given that the average temperature at the site under study is high, which would provide a higher level of comfort for the workers.

Ergonomics and safety is a relatively new science, a topic that has been discussed for around 10 years. The most appropriate solution to the issue of safety and, above all, ergonomics is to exchange old vehicles for new and used ones, which would be possible through public policies that provide lines of finance. Drivers need to be aware of the depreciation of their vehicles, and so save up so that at the end of the economic threshold they can exchange their vehicle. According to Machado (2000), the fleet should be renewed at least every 10 years.

Ergonomics is seen as a science that can act to prevent accidents. In the event of the occurrence of this pernicious event in the system (the accident), ergonomics contributes through ***design*** with passive safety elements such as seat belts, motorbike helmets, padded dashboards, air bags and road defences, with the aim of minimising the number of deaths and injuries (PETZHOLD, 2011).

2.5.2. Transport problems

Still from the perspective of each driver's opinion, they were asked what the biggest problem in forestry road haulage was. The problems cited were diverse.

The problems related to the load were as follows: the majority reported that the load had come loose from the vehicle, due to the cables breaking, as well as reports of a problem on the track that caused the vehicle to tip over. Most of them agreed that the problem starts in the operating area, when the load leaves the yard crooked, i.e. poorly arranged, which means that there is a risk of the load falling or the vehicle overturning if it is travelling on uneven ground.

For 38% the biggest problem is the unmaintained roads, 14% is the poor freight, 10% enforcement and/or the problem of driving without a note, 8% bad bridges/poorly maintained bridges, 8% delay in receiving a note, 6% lack of a note, 4% when the load is opened it delays the journey, 4% flat tyres on the road, 4% overweight fines. Dangerous cargo, bureaucracy, long distances, obstacles with environmental laws, loading and unloading, because it's a dangerous operation, delay in unloading, cargo without documents, delay in getting the note, a lot of traffic on the roads, inspection.

Reports from hauliers on the biggest problems in forestry freight transport by road:

> **"...In the note there is a route from the bush to the city, the logging company pays a "fine" if it goes off the route. In the bush you don't know how heavy the lorry is, so it comes in overloaded and gets fined. Bureaucracy is a problem, without a note the transporters are stuck; if the system is down (the internet goes down) the note doesn't come out and delays the journey; it takes a long time for the note to come out; there are conflicts between the environmental agencies, between Sema and Ibama, where Sema releases and Ibama ends up "barring", in relation to the roads there are some stretches where the roads are narrow, mainly over bridges. The time it takes to get the note: To get out of the bush you have to wait for the note, and sometimes it takes up to 3 days and the lorry driver has to wait all that time in the bush."**

As presented in the work by Melo (2009), the biggest difficulty encountered by drivers when transporting raw materials, totalling 100%, is the roads and bridges. This high percentage is justified by the inconvenience caused to these workers, which could be alleviated with their maintenance and upkeep. These inconveniences mean hours lost waiting on the roads and costs for tyres, parts and mechanics in general. The drivers' reports were similar as to why they considered the cargo to be dangerous:

> "The load can come off. You have to be careful and attentive, because the load can overflow, it's liable to roll, because it's cylindrical, so the load becomes unpredictable, the cables loosen during the journey, if the logs are badly fitted, you have to be careful on the road and when lashing, loading and unloading, many people have died as a result of incidents with the load... Because it's a high and heavy load, you have to be more careful in the city, it's unpredictable, easy to get into trouble if the log is badly adjusted... Risk due to excess weight, risk of tipping over, cables snapping, danger for those without experience."

When questioned, the drivers made a number of observations, some of them adding that there are many lorries that use "protective covers" in transport in Mato Grosso, while in other states like Rondônia they don't use them. When they leave the forest, they "pull the cables" at least three times to get to the logging company, over a 200-kilometre journey. As they drive slowly, they say they don't compensate by wearing seatbelts and only use them on federal highways. The greatest risk of an accident is when the driver is loosening the cables while the shovel is removing the logs, because the risk of one of the logs rolling over is great due to the swaying of the truck (it moves the whole trailer). The height of the staves is now higher, which makes them safer. The Federal Police require the stanchions, which are also called "life-saving pins" and "giant" panels, which are important for safety. The logs are positioned in the shape of a pyramid. It is a mistake to place the thinner logs under the thicker ones. Some have admitted to being fined by the Federal Police because the logs exceeded the maximum height and length allowed.

One driver argues about excess weight, which he says is a driver's excuse for not being aware of how much weight is in the vehicle, where according to him if the driver wanted to try not to be overweight, this would be possible:

> "Most of the time the lorry is overloaded, where, for example, 50 tonnes would only be 2 stacks of logs. You can get a sense of the weight of the load in the forest, ***because each m^3 is on average 1,300kg, but it doesn't pay to carry just 50 tonnes, so the driver carries twice as much. Some of the species they were transporting: itaúba, cedrinho, canelão, garapeira, champanhe, massaranduba. A load of wood is valued at around R$500,000.00. Freight is bad, it should be 30-40% higher, diesel is very expensive, maintenance is also expensive."***

4.2. Vehicles

4.2.1. Main activity

When asked about the vehicle's annual activity, i.e. whether the activity is solely forestry or whether during the year, during periods of low or no logging, the vehicle also transports agricultural cargo, it was found that 54 per cent of the vehicles alternate the harvest with the transport of raw timber in logs during the year.

4.2.2. Year of vehicle manufacture

Figure 14 shows the year of manufacture of the vehicles used in the municipality of Sinop-MT.

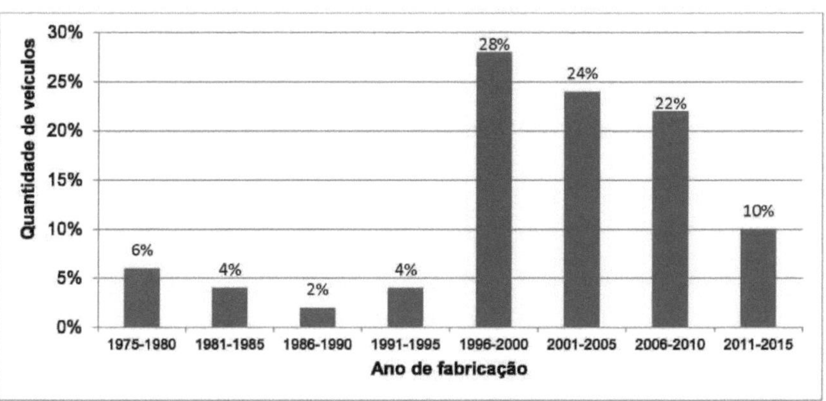

Figure 14 - Distribution of vehicles by year of manufacture used to transport loads of raw timber in the municipality of Sinop-MT.

According to the results obtained, the average age of the fleet was 12 years, with 28 per cent of vehicles in the 1996-2000 class, followed by the 2001-2005 class with 24 per cent of the total sampled. This research showed that the year of manufacture varies, with new, used and old vehicles. Of the entire fleet, the oldest was 35 years old. As age and kilometres driven are related, the survey found that the average total kilometres driven was 258,482.60km.

It can therefore be said that the municipality's fleet is relatively old, given that ideally it should be no older than 10 years (MACHADO, 2000). However, 66% of those interviewed said that the municipality had a medium-quality fleet for transporting raw timber, given the state of repair of the vehicles. Melo (2009) indicates that for the municipality of Alta Floresta in the north of Mato Grosso, in relation to the age of the fleet, an average of 12 years was also found, with 36.66% of the vehicles being of this average age, which according to Machado (2000) is also an ageing fleet, however 76.66% of those interviewed said that the municipality had a good quality fleet for the purpose for which it was intended, compared to other regions in the Legal Amazon.

Only government investment, such as lines of credit and financing with more accessible interest rates for workers, could help to change this fleet and provide greater satisfaction and well-being for drivers, thus helping to increase productivity.

4.2.3. Vehicle make

Figure 15 shows the makes of the vehicles used to transport loads of roundwood in the municipality of Sinop-MT.

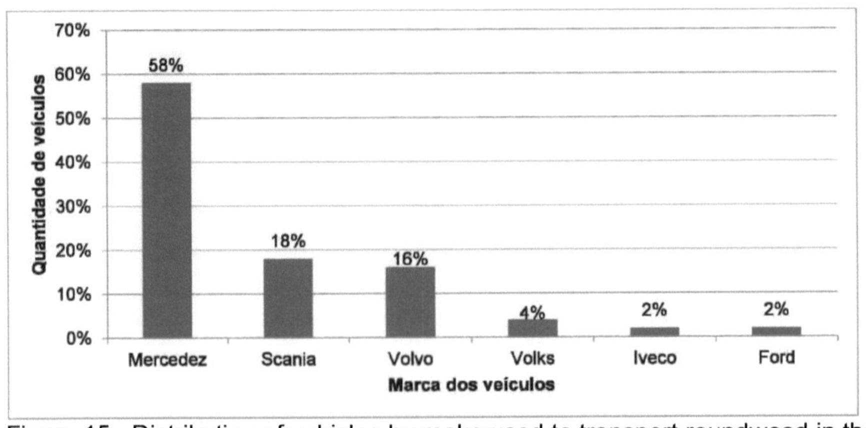

Figure 15 - Distribution of vehicles by make used to transport roundwood in the municipality of Sinop-MT.

Among the vehicle makes, the most used in forestry transport in the municipality was Mercedes Benz, which accounted for 58% of the vehicles sampled. Melo (2009) confirms the 60 per cent preference for this brand of vehicle, because in the opinion of those interviewed, this is because this brand has the greatest variability of models, which when added to the Julieta combination, provide evidence of optimum operational performance.

4.2.4. Vehicle combinations

Figure 16 shows the combined vehicle combinations of the vehicles used in the transport of roundwood in the municipality of Sinop-MT and the region.

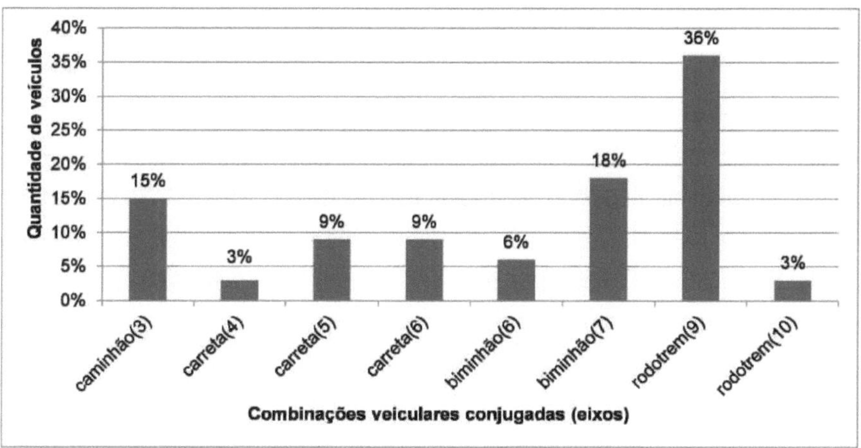

Figure 16 - Vehicle combinations used to transport logs.

Based on the graph, the vehicle combinations used were analysed. This analysis was carried out in accordance with the photographic archive, however many drivers did not allow the image to be taken, and some of the images obtained do not show the side view of the vehicle, but only the front and rear. As a result, 33 images capable of distinguishing the category of vehicle were recorded. According to the graph, 36 per cent of the CVCs were 9-axle road trains, followed by 18 per cent 7-axle bimini trucks.

The "Romeo and Juliet" combination, i.e. 6-axle bimini trucks, which was very common in earlier times, was expected to be found in road forestry transport, but only 6% of the fleet was seen. This is an alarming indication that

combinations are getting bigger and bigger. The lesser use of tractor-trailers is explained by studies which have found that articulated and combined vehicles perform better over distances of more than 135 kilometres and with a minimum load volume of 66 stereos of wood. Single lorries, despite having lower fuel consumption than other vehicle combinations, have limitations in terms of load capacity and, consequently, in terms of transporting timber over long distances. In order to obtain better results in rationalising road forestry transport operations, it is necessary to know the main factors that determine the optimum performance of vehicles (MACHADO, 1991).

In general, it can be seen that the use of rodotrens, tritrens and bitrens means fewer vehicles on the roads, generating less fuel wastage, and should therefore be a reason for prioritising interests and procedures in forestry transport logistics operations, as well as in other important segments of the Brazilian economy. To illustrate the optimisation of loads and associated operational gains, the following are the average volumes transported for these compositions: bitrens - 39 tonnes/74 sterile cubic metres/46 compact cubic metres; rodotrens - 53 tonnes; and tritrens - 51 tonnes/110 sterile cubic metres/70 compact cubic metres (SEO, 2010).

Marchiori (2009) describes the mandatory signage with a plate on the rear of the last trailer, as determined by CONTRAN Resolution No. 211 of 13 November 2006. Art. 1° Cargo Vehicle Combinations (CVC) with more than two units, including the tractor unit, with a total gross weight of more than 57 tonnes or a total length of more than 19.80 metres, may only be driven with a Special Transit Authorisation (AET). And obeying the special signage for combinations of cargo vehicles - CVC rear plate (for combinations with a length exceeding 19.80m) and side reflective plates.

4.2.5. Power classes

Figure 17 shows the power classes of the vehicles used to transport logs in the municipality of Sinop-MT.

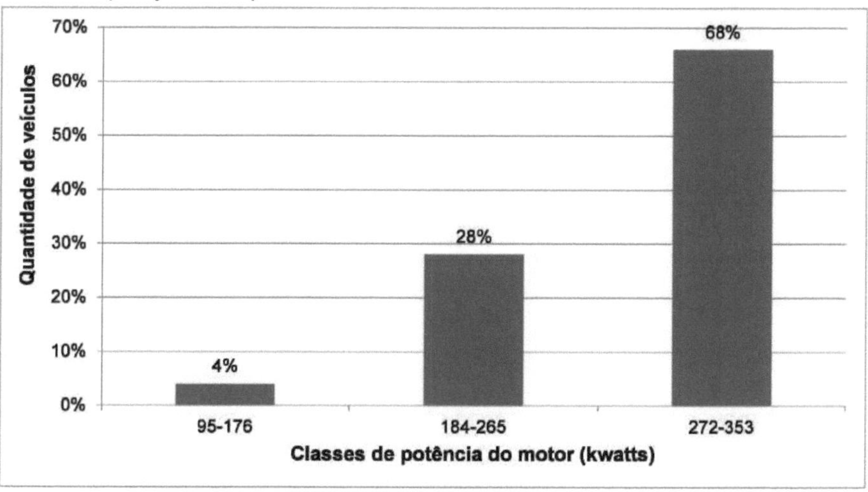

Figure 17 - Distribution of vehicles by engine power (hp) used to transport roundwood in the municipality of Sinop-MT.

Variations in engine power were noted, as this is a factor that influences the weight of the loads transported. It was observed that 68% of the engines in use have a power of 272 to 353 kw, which is equivalent to 370 to 480 hp, demonstrating a good power capacity compared to Melo (2009) who in his survey pointed out that 80% of the engines in use have power in the 320-400 hp class, demonstrating the need for greater power due to the use of larger trains to make transport viable. In 2005, 40% of the vehicles used had power between 360 and 380 hp in the same municipality of Alta Floresta (PEREIRA, 2005).

4.2.6. Load capacity

Figure 18 shows the load capacity classes in tonnes of the vehicles used to transport logs in the municipality of Sinop-MT.

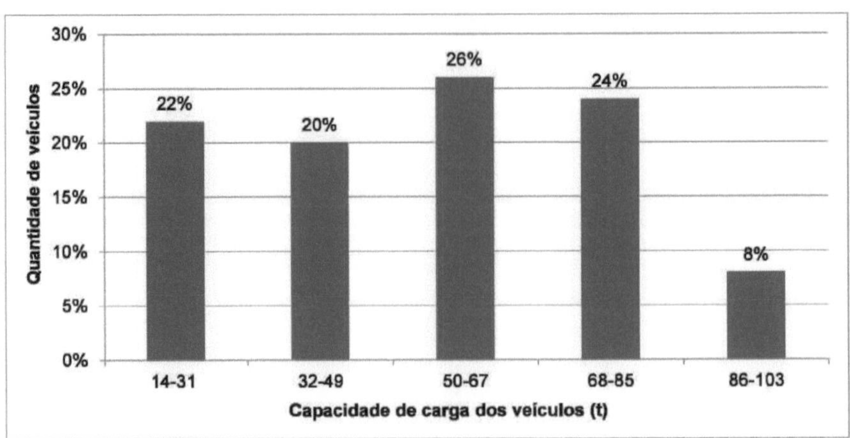

Figure 18 - Distribution of vehicles by load capacity (tonnes) used to transport logs in the municipality of Sinop-MT.

The graph shows that 26% of the vehicles analysed were in the 50 to 67 t load class, followed by vehicles with a load capacity of 68 to 85. For Melo (2009), 33.33% of the vehicles sampled had a load capacity of between 36 and 50 tonnes. Followed by vehicles with a capacity of 51 to 601 tonnes.

Figure 19 shows the load capacities of the vehicles (m^3) used to transport logs in the municipality of Sinop-MT.

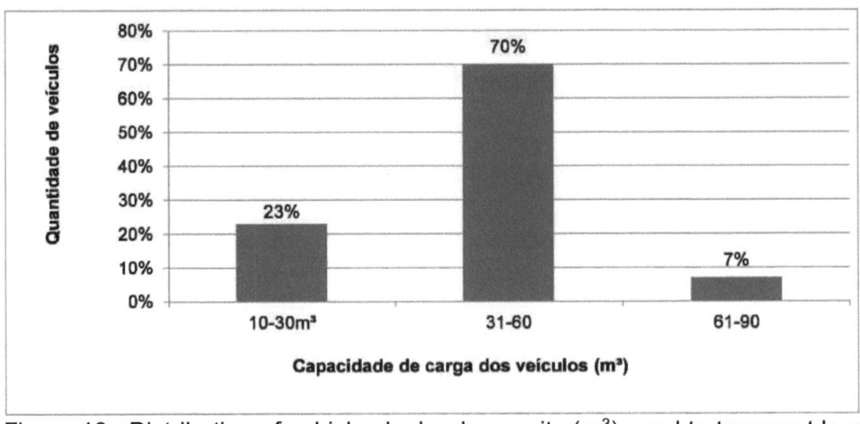

Figure 19 - Distribution of vehicles by load capacity (m^3) used to transport logs in the municipality of Sinop-MT.

With regard to load capacity in cubic metres, 70% of the vehicles had a load capacity of between 31 and $60m^3$. In Alta Floresta, 56.66% of the vehicles were able to transport between 36 and 50 $metres^3$ (MELO, 2009). Load capacity is influenced by the power of the vehicle carrying out the transport, as the greater the power, the greater the amount of raw material transported. This demonstrates the need for increasingly larger compositions for transport in the municipality.

4.2.7. Property rights

Figure 20 shows the distribution of vehicles in relation to the outsourcing of transport in the municipality of Sinop-MT.

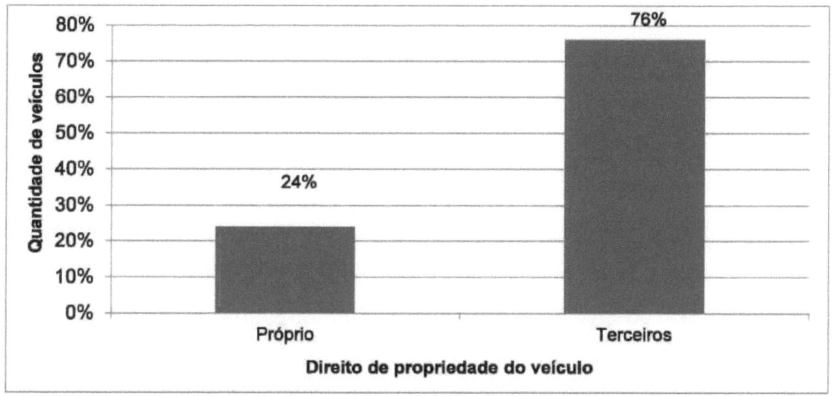

Figure 20 - Distribution of vehicles in relation to the outsourcing of roundwood transport in the municipality of Sinop-MT.

As can be seen in the graph, the majority of the vehicles were owned by third parties, corresponding to 76% of the total sample, i.e. outsourced means that the vehicles are owned by transport companies, by the logging companies themselves or by a third party. Most of the vehicles belong to companies specialising in forestry transport and 24% of the vehicles are owned by the driver.

According to a recent discussion within the International Labour Organisation (ILO, 2003), truck drivers in many countries have been greatly affected by the organisational changes implemented by companies with the aim of increasing their level of competitiveness. This has led to a shift from workers with permanent employment relationships, absorbed by haulage companies, to the situation of "outsourced" workers, or self-employed truck drivers, constituted in the legal form of autonomous or entrepreneurs, who must have their own lorry, or even rent it, carrying out practically the same services as those they used to.

previously carried out as an employee. (GOLDIN and FELDMAN, 1999; HYDE, 2000).

As a result, in practically all cases the driver is legally separated from the company, but continues to perform the same job in a position of effective dependence on the employer. The difference now is that the driver is autonomous and must bear the costs of maintaining the vehicle without any labour protection (CHAHAD and CACCIAMALI, 2005).

Unlike the present study, where the aim was to find out whether the fleet was owned by the drivers, the aim of Melo's study (2009) was to find out whether the vehicles used were owned by the timber company itself or by third parties, i.e. Melo's term "third parties" is equivalent to the term "own" in the present study. As a result, it was shown that the timber transport activities used by the timber industries assessed are highly outsourced. As proposed by Melo et al. (2009), 83.33% of the transport is carried out entirely by agents with no employment relationship with the timber companies (autonomous drivers and carriers) and only 16.67% are entirely the timber companies' own vehicles.

In 2005, the rate of outsourced services was 30 per cent, while those carried out by the company itself was 50 per cent (PEREIRA, 2005). Outsourcing has therefore become a current trend for companies of various sizes and segments in order to survive highly competitive scenarios (MACHADO, 2000).

Companies that use road transport to deliver goods always want to know what the lowest cost is between their own fleet and an outsourced one. Many of these companies outsource, believing it to be the best option for minimising their costs. Others prefer to own the fleet, betting on the quality of the distribution service (SILVA, 2008).

According to Fleury et al. (2004), freight usually absorbs approximately 60 per cent of logistics costs. For this reason, companies are keen to minimise these costs in the transport sector, as they know that they represent around

20% of the total cost.

According to Figueiredo et. al. (2006), there are many transport service providers on the market, such as logistics operators and autonomous transport, with attractive costs. Bailou (2001) states that transport is a determining factor in logistics costs for most companies.

4.2.8. Maintenance and state of repair

Figure 21 shows how many drivers carried out preventive maintenance.

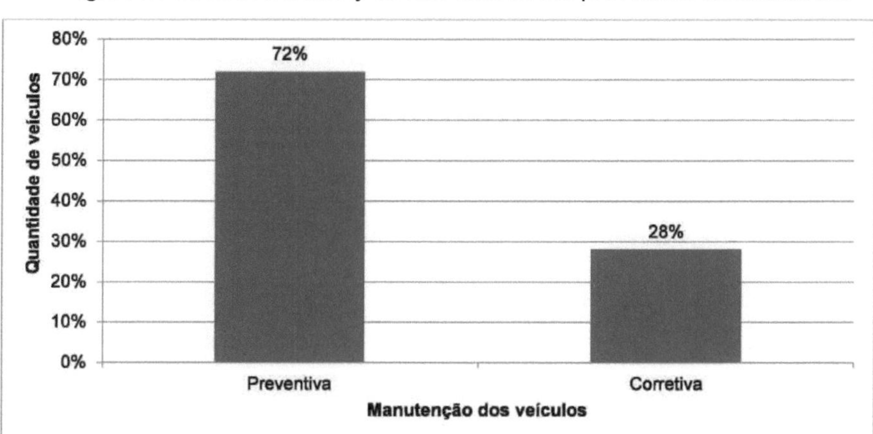

Figure 21 - Distribution of the type of maintenance carried out on log transport vehicles in the municipality of Sinop-MT.

72 per cent said they carry out preventive maintenance, which is essential to avoid higher costs in the future.

Agreeing with the importance of maintenance, Moro (2007) gives an idea of the need to establish a maintenance programme, since if machines and equipment are defective and/or stopped, the damage will be inevitable, causing: reduced or interrupted production; delays in deliveries; financial losses; increased costs; customer dissatisfaction and loss of market share.

The concept of corrective maintenance is a set of procedures that are applied to totally or partially damaged equipment, with the aim of getting it back to work in the shortest possible time and at the lowest possible cost. Preventive maintenance, on the other hand, involves establishing periodic stops to allow for scheduled repairs, thus ensuring that the machine works perfectly for a predetermined period of time (MORO, 2007).Figure 22 shows the drivers' opinions on the state of conservation of the fleet used to transport raw timber in the northern region of Mato Grosso.

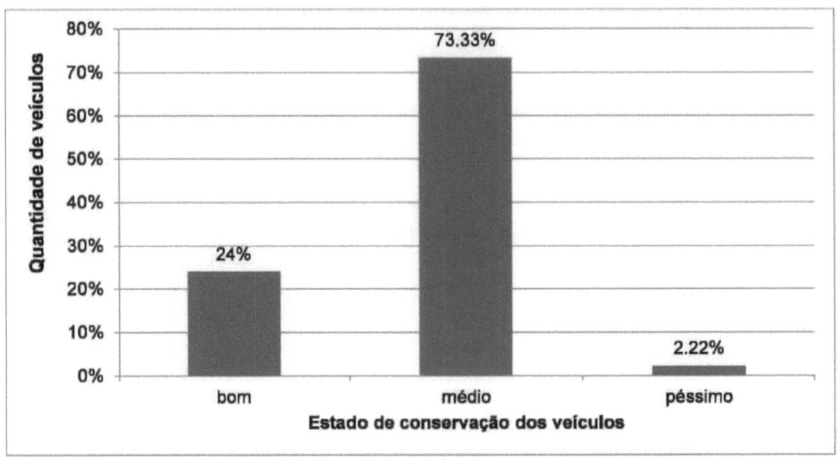

Figure 22 - Drivers' perception of the state of repair of vehicles.

73.33% of the drivers interviewed considered the regional log transport fleet to be in an average state of repair. However, one of the drivers said it was terrible, citing the example of a nearby town called Feliz Natal, where according to him the lorries are scrapped, and he also said that the further you go into the interior of the state, the worse the condition of the vehicles.

Some of the aspects that contribute to the poor performance of road transport in Brazil are: the high average age of the fleet - around 1.5 million vehicles - which are, on average, almost 16 years old (BARTHOLOMEU, 2008). According to Machado (2000), the age of the fleet should not exceed 10 years. In this study, the overall average age was 12 years.

Another interesting point in the survey was that some of the interviewees noticed that trucks were being increasingly scrapped in other states. One of the interviewees added that he had worked in the forestry sector in the state of Rondônia and that the fleet there was older than in our state. "In our region the lorries are well maintained, but in the cities of the interior of Rondônia they are scrapped."

4.2.9. Developed speeds

In this study, the speeds of log transport vehicles on tarmac and dirt roads were obtained.

Figure 23 shows the vehicle speeds on the tarmac.

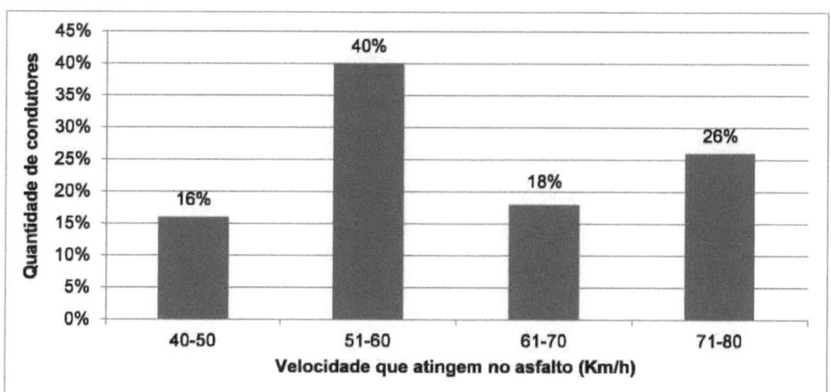

Figure 23 - Distribution of drivers with regard to the speed they reach on the tarmac while transporting the load.

According to the results, we found that 40 per cent of drivers drove their loaded vehicles at an average speed of 51 to 60km/h. Freight transport in Brazil is highly dependent on road transport. Consequently, the performance of this mode of transport ends up suffering. Compared to the performance observed in the USA, for example, the productivity of Brazilian road transport is 22% lower, while energy consumption and carbon monoxide emissions (in gCO/t.km) are 29% and 2.6 times higher than in the USA (CEL et al., 2002).

Some of the aspects that contribute to the poor performance of road transport in Brazil are: the high average age of the fleet (vehicles - around 1.5 million - are almost 16 years old on average) and an insufficient supply of road transport infrastructure, both in terms of length and quality of roads. The results obtained confirm the hypothesis that routes with different infrastructure conditions result in different travel costs, related to fuel consumption, journey time and vehicle maintenance costs. Therefore, routes in better condition result in greater economic and environmental benefits (BARTHOLOMEU, 2008).

A study carried out by technicians from the National Highways Department

(DNER) and the Brazilian Transport Planning Company (Geipot) shows that a degraded road represents a 58 per cent increase in fuel consumption, 38 per cent in vehicle maintenance costs, 50 per cent in the accident rate and up to 100 per cent in travel time (Revista CNT, 2001).

According to Terra (2012), the Mercedes-Benz Actros 4844 K model has the capacity to pull 123 tonnes, with an engine that generates 435 horsepower, at a top speed of 79 km/h. By way of comparison, one of the models also discussed in the interviews was the 400 horsepower Mercedez-Benz Axor 2340, loaded with an average of 70 tonnes of logs, with an average speed on paved roads of 80 km/h. These figures are similar to all the drivers of this truck model.

Figure 24 shows the speed data for dirt roads.

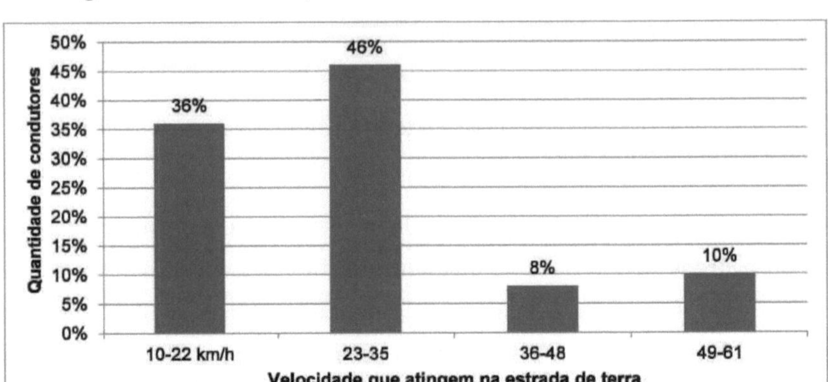

Figure 24 - Distribution of drivers according to the speed they reach on the dirt road while transporting the load.

According to the graph, 46% of the vehicles when loaded and subjected to unpaved roads have a speed of between 23 and 35km/h. According to 35 drivers, that is, 70% of the vehicles fell into the extra heavy power class, and 56% of the drivers said they were carrying loads of between 50 and 103 tonnes. Because the vehicle is loaded and, as reported, usually travelling overweight, coupled with poorly maintained roads, the speed is reduced.

Among the problems with the roads, the most frequently cited were: poorly maintained and narrow bridges, bumpy roads, lack of maintenance, lack of shoulders, uneven roads, difficult access to food along the way, in short, in general it was noted that there is a need for constant maintenance, which has not been happening.

4.2.10. Distance travelled

Another important factor in timber transport is the distance travelled from the logging area to the final destination for processing. The data on log transport distances, from the place of origin of the cargo to the logging yards in the municipality of Sinop, is shown in Figure 25.

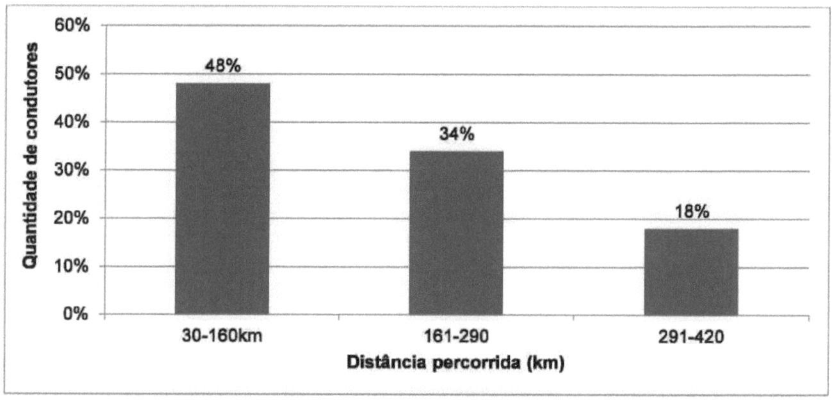

Figure 25 - Distances travelled from the logging area to the timber yard.

According to the data in the graph, the minimum distance travelled from the logging area to the timber yard was 60km and the maximum was 380km, with an overall average of 195km.

It was observed that 48% travelled between 30 and 160 km. According to Pereira (2005), the transport distance between the raw material in the storage yard and the industry in 2005 was an average of 100 km for the municipality of Alta Floresta-MT. It is therefore assumed that the increase in this distance is due to the scarcity of raw materials close to the municipality.

Melo (2009) found that 86.66% travelled between 50 and 200km.

To carry out this operation, the average time they reported taking to bring in the load was just over 10 hours.

The average distance was 117 kilometres between the logged forests and the processing industries. In regions where river transport predominates, the distance was greater: up to 400 kilometres, as in Pará. In the new logging frontiers, the distance was shorter and fluctuated around 81 kilometres (PEREIRA, 2010).

4.3. Origin of cargo

Figure 26 shows the municipalities of origin of the log cargo destined for the municipality of Sinop.

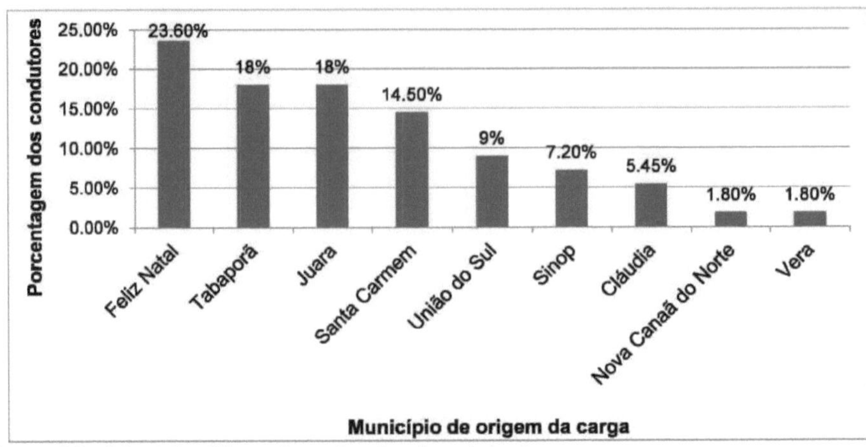

Figure 26 - Municipalities where log loads originate.

This graph shows the most frequent logging destinations, i.e. the municipalities where the cargo originated, where 23.6% came from Feliz Natal, 18% for each of the two municipalities.

respective municipalities: Tabaporã and Juara, with 14.5 per cent from Santa Carmem, followed by União do Sul with 9 per cent and the municipality of Sinop itself with 7.2 per cent.

The distances between the municipalities where the log loads originate and the municipality of Sinop are shown in Table 6.

Table 6 - Distance between Sinop and the municipalities where the wood originates.

Municipality of Origin	Distance (Km)
Merry Christmas	121
Tabaporã	170
Juara	291
Santa Carmen	38
Southern Union	160
Cláudia	80
Nova Canaã do Norte	230
Vera	72

SOURCE: Distance between cities.

4.4. Road and motorway traffic conditions

4.4.1. Motorways

Most of the drivers pointed out that the flow of traffic on our roads is significant, especially during the harvest period, and that conditions are precarious on some stretches due to the lack of shoulders and signposting.

Drivers were asked about the traffic conditions on federal, state and municipal roads. The data obtained is shown in Figure 27.

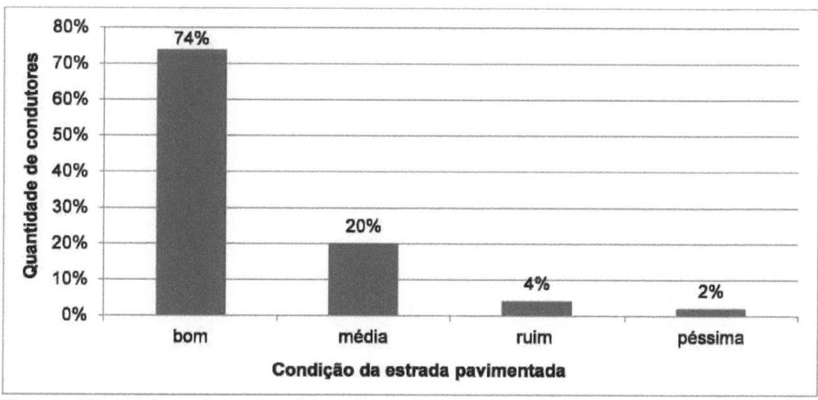

Figure 27 - Drivers' perception of road traffic conditions.

The graph shows that 74% found the roads to be in a good state of repair, which is positive, since according to Nepstad et al. (2000) there are many sawmills along the BR-163 that benefit from paving. With good roads there is a reduction in transport costs, especially in the rainy season, when there is an 80 per cent reduction. This makes the region's timber more competitive on both the southern Brazilian and international markets.

Of the total supply of roads (around 1.6 million kilometres), only 12% are paved (CNT Statistical Bulletin, 2005b). Of the paved state portion analysed by the Road Survey (CNT, 2005a), more than 59,000 km, or around 72%, have pavements classified as "Poor", "Bad" or "Bad". Well-maintained roads actually result in lower transport costs, due to the reduction in fuel consumption, journey time and truck maintenance costs (BARTHOLOMEU, 2008).

4.4.2. Roads

Figure 28 shows drivers' opinions on road conditions.

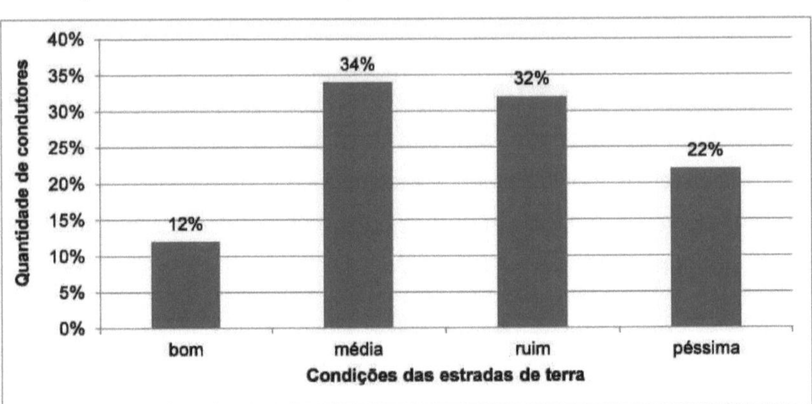

Figure 28 - Drivers' perception of road conditions.

The graph shows the dissatisfaction with road conditions, where the sum of those who consider the roads to be between bad and terrible is 54%. The lack of paving on side roads hampers the transport of grains planted in the central region of the state of Mato Grosso, making the product less competitive compared to other states with better infrastructure (WINTERS, 2013).

In the Amazon, around 36 per cent of roundwood was transported on "gravelled" roads, i.e. those that generally have better road conditions than "ungravelled" roads, as they have better drainage, levelling and conservation, another 30 per cent on ungravelled roads, 19 per cent on asphalt and 15 per cent on waterways (ferries or rafts) (PEREIRA, 2010).

CHAPTER 5

FINAL CONSIDERATIONS

From the results obtained on the perception of drivers transporting raw timber logs by road freight vehicle in the Legal Amazon, it was possible to conclude that:

Their long time in the profession was explained by their love of working as drivers. Most of the drivers had a BMI outside the normal range, which was proven by the fact that they were overweight. The quality of life of these drivers was jeopardised by the fact that most of them drove in the early hours of the morning. Another risk to health and safety was the fact that they didn't wear their seatbelts regularly.

- The fleet varied greatly in age, from 35-year-old vehicles to new ones. The lack of safety was also noted in the height of the loads, as they were outside the standards required by CONTRAN. The destination of the wood loads transported by the drivers interviewed was the municipality of Sinop. The raw materials came from the towns around Sinop, the nearest being Santa Carmem (38 kilometres) and the furthest Juara (291 kilometres).

- Only government investment, such as credit lines and loans with interest rates that are more accessible to workers, could help to replace this fleet and provide greater satisfaction and well-being for drivers, thus helping to increase productivity through more powerful vehicles.

BIBLIOGRAPHICAL REFERENCES

ALVAREZ, B. R. **Health-related quality of life of workers.** Dissertation (master's degree) - Federal University of Santa Catarina, Technology Centre. 2012. Available at: http://repositorio.ufsc.br/handle. Accessed on 07 February 2013.

AMARAL, K. M. **Rounding the world:** the labour relations, experiences and lives of truck drivers. X National Meeting of Oral History. Testimonies: Historical and Political. Federal University of Pernambuco (UFPE) Centre for Philosophy and Human Sciences. Recife, 26 to 30 April 2010. Available at: http://www.encontro2010.historiaoral.org.br/resources/anais/. Accessed on 05 February 2013.

AUSTROADS - AUSTROADS NATIONAL OFFICE. **Road Safety Audit.** Sydney, 1994.

Auto Esporte. Rio de Janeiro: TV Globo, 28 March 2011. TV programme.

BALLOU, R. H. **Supply Chain Management.** 4ª Ed. Bookman. Porto Alegre, 2001.

BARROS, R. P. et al. **An analysis of the main causes of the recent fall in Brazilian income inequality.** Económica, 8(1): 117-147. 2006.

BARTHOLOMEU, D. B.; CAIXETA FILHO, J. V. Economic and environmental impacts of the state of conservation of Brazilian motorways: a case study. **Revista** de Economia e Sociologia Rural. V. 46, n. 3, p. 703-738, 2008.

BATISTA, M. A.; SILVA, F. A. **Level of health of lorry drivers travelling along the BR 040, based on data obtained during the VI Federal Highway Command, in the city of Brasília - DF.** A Case Study. Brasília, p.02-12, 2005.
BIODIESELBR. Transport in Brazil. **BiodieselBR Magazine. 18th** August 2006. Curitiba- PR. Available at: http://www.biodieselbr.com/destaques/analise2/transportes-no-brasil.htm. Accessed on 25 May 2012.

BRACELPA. Forestry Sector. **Strategic thinking.** 2011. Available at: http://www.holtz. com.br/index.php?option=com_content&task=view& id=20&Itemid=7. Accessed on 25 May 2012.

BRAZIL. Ministry of Justice, Denatran. **Municipalisation of Traffic:** roadmap for implementation. Brasília, 2000.

BRAZIL. MINISTRY OF PLANNING, BUDGET AND MANAGEMENT - IPEA. **Social and economic impacts of traffic accidents in urban agglomerations.** Brasília, 2003.

BRAZIL. Law No. 9.503/1997, of 23 September 1997. DENATRAN - National Traffic Department, Brasília, 23 September 1997.

CÂNDIDO, N. **The importance of seat belts.** 2009. Available at: http://itabi.infonet.com.br/trocandoideias/?p=298. Accessed on 05 February 2013.

CHAHAD, J. P. Z. CACCIAMALI, M. C. Structural transformations in the road transport sector and the reorganisation of the truck driver labour market in Brazil. **Journal of ABET - Brazilian Association of Labour Studies.** Year 5, v. 2, n.10, 2005.

CENTRE FOR LOGISTICS STUDIES (CEL); NATIONAL TRANSPORT CONFEDERATION (CNT). **Freight transport in Brazil:** threats and opportunities for the country's development. Rio de Janeiro, September 2002. Available at: http://www.coppe.ufrj.br. Accessed on 18 June 2012.

CHAFFIN, D. B.; ANDERSSON, G.B.J. **Occupational biomechanics.** New York: John Wilwy & Sons, 1990.
COELHO, L. C. **Types of lorries, sizes and capacities.** Logística descomplicada. November 2010. Available at: http://www.logisticadescomplicada.com. Accessed on 20 April 2011.

WOOD HARVESTING. **Schematic of combined vehicle compositions.** Authors: Timber Harvesting Team. Technical material.
Curitiba, 2012. Available at http://www.colheitademadeira.com.br/. Accessed on 23 January 2013.

CNT - NATIONAL TRANSPORT CONFEDERATION. **Situation of Freight Transport in Brazil.** 2002. Available at http:/www.cnt.org.br/download/pesquisa/cnt-coppead-cargas.pdf . Accessed on 22 November 2012.

CNT - NATIONAL TRANSPORT CONFEDERATION. Statistical bulletin. Brasília, Dec. 2005b. Available at: http://www.cnt.org.br/cnt/downloads/becnt/becnt_122005.pdf. Accessed on 15 May 2012.

CNT - NATIONAL TRANSPORT CONFEDERATION. **Road research 2005:** management report. Brasília, 2005a. Available at: http://www.cnt.org.br. Accessed on 15 May 2012.

CNT - NATIONAL TRANSPORT CONFEDERATION. CNT **Magazine**. Brasilia, 2001. Various issues. Available at: http://www.cnt.org.br. Accessed on 15 May 2012.

CONTRAN- NATIONAL TRANSIT COUNCIL. Resolution N°188 of 25 January 2006. Available at: http://www.pr.gov.br/mtm/legislacao/resolucoes/resolucao188_06.htm.

Accessed on 19 March 2012.

CORREIA, L.; SILVA, M. R. Anthropometric profile of senior basketball players from the Viana do Castelo Basketball Association. **Revista** Faculdade de Ciências da Saúde. Porto: University of Fernando Pessoa. Portugal. 2009.

CZEPIELEWSKI, M. A. **Obesity.** ABC da saúde medical information ltda. 2001. Available at:www.abcdasaude.com.br/artigo.php7303 . Accessed on 20 February 2013.

DISTANCE BETWEEN CITIES. Available at: http://www.entrecidadesdistancia.com.br/. Accessed on 28 February 2013.

ENGLAND, L. The role of accident investigation in road safety. **Ergonomics.** N 24, v 6, p 409 - 422, 1981.

FALEIRO, N. S. **The research project.** Research methodology and technique. Univates. 2011.

FELIPPE, L. F. et al. Prevalence of Postural Alterations and Musculoskeletal Pain in Truck Drivers. **Revista** Movimenta. ISSN: 1984- 4298. Vol 5. N 2. 2012.

FIEDLER, N. C. **Ergonomic assessment of machines used in wood harvesting.** 1995. 126 f. Master's thesis in Forestry Science. Federal University of Viçosa-MG. Viçosa, 1995.

FIGUEREDO, K. F.; FLEURY, P. F.; WANKE P. **Logística Empresarial.** 1ª Ed. Atlas. São Paulo, 2006.

FLEURY, P. F.; WANKE, P.; FIGUEIREDO, K. F. Centro de estudos em logística. **Business logistics:** the Brazilian perspective. Organisation. São Paulo: Atlas, 2004.

GOLDIN, A. and FELDMAN, S. **Worker Protection:** Argentine Report. Responses to the questionnaire, Country Studies: Argentina, ILO, Geneve.1999.

GUIMARÃES, R. R. de M.; FÍGOLI, M. G. B.; OLIVEIRA, A. M. H. C. de. **Staying in Precariousness and Decent Work:** A multi-state model for transitions according to quality of occupation for Metropolitan Brazil (2003-2007). CEDEPLAR/UFMG, 2010.

GUIMIER, D. Y.; WILLBURN, G. V. **Logging with heavy-lift airships.** Vancouver: FERIC, Technical Report No. TR-58, 1984. 115 p.

HYDE, A. **Classification of U.S. Working People and Its Impacts on Workers Protection.** Country Studies: USA, ILO, Geneve. 2000.

ICV - CENTRE OF LIFE INSTITUTE. **The Forestry Sector in the Portal Territory of the Amazon, north of Mato Grosso:** Current situation and prospects. 2006. Available at: http://antigo.yikatuxingu.org.br/arquivos/projetos/72/diagnostico_florestal_portal.pdf Accessed on 21 September 2011.

IIDA, I. **Ergonomics:** design and production. 9ª ed. São Paulo: Edgard Blucher, 465p. 2003.

ILO. **The Scope of the Employment Relationship.** 91st Session, Report V, Fifth Item on the Agenda, International Labour Office, Geneve.2003.

IPEA- INSTITUTE OF APPLIED ECONOMIC RESEARCH. Economic climate report. **Technical report,** 2007.

KILESSE, R. et al. **Evaluation of ergonomic factors in the workstations of lorry drivers used in agriculture.** 2006.
Available at:
http://br.monografias.com/trabalhos901/fatores-ergonomicos-trabalhoomotoristas/fatores-ergonomicos-trabalho-motoristas.shtml. Accessed on 12 April 2012.

MACHADO, C. C.; LOPES, E. da S.; BIRRO, M. H. B. **Basic elements of road forestry transport.** Viçosa-MG. Federal University of Viçosa, Ed. 2000.

MACHADO, C. C. **Forest Harvesting.** Viçosa, MG. UFV, 2008.

MACHADO, C. C.; LOPES, E. da S.; BIRRO, M. H. B. **Forestry Road Transport.** Viçosa-MG. Ed. UFV, p.64. 2009.

MACHADO, C. C.; SOUZA, A. P. de; LEITE, A. M. P. **Analysis of the performance of different road forestry transport vehicles.** Revista Árvore, Viçosa, v. 15, n. 1, p. 67-81, 1991.

MACHADO, C. C.; SOUZA, A. P. **Safety when working with chainsaws.** Viçosa, MG: Federal University of Viçosa, p.10. 1980.

MAGRI, C. F. et al. Perspectives on the agricultural production sector along the Cuiabá-Santarém motorway. Anhanguera **Magazine**. Year 6, series 1, p. 9-34. Goiânia. 2005.

MANZATTO, L. **Qualidade de Vida no Trabalho:** Avaliação Quali/Quanti de Drivistas de uma Empresa de Transporte Rodoviário de Cargas.
Dissertation presented to the
Examiner for the Postgraduate Programme in Physical Education at UNIMEP. PIRACICABA. 2012.

MARCHIORI, A. Forestry transport. Document type: Occupational health and safety procedure. INTERNATIONAL PAPER. 2009.

MARTINS, R. S.; CAIXETA FILHO, J. V. Subsidies for Decision-Making on the Choice of Modality for Transport Planning in the State of Paraná. **Journal of** Contemporary Administration. V. 3, n. 2, p. 75-96. 1999.

MCARDLE, W. et.al. **Exercise physiology, energy, nutrition and human performance.** 3ª ed. Rio de Janeiro: Guanabara Koogan. 2002.

MELO, C. R. D. B. **Survey of vehicles used in forestry transport in the region of Alta Floresta-MT.** Course Conclusion Paper (Graduation) - Faculty of Forestry Engineering, Agro-Environmental Sciences Programme, Mato Grosso State University, Alta Floresta. 2009.

MINETTE, L. J. **Analysis of operational and ergonomic factors in forestry felling operations with chainsaws.** 211f. Thesis. PhD in Forestry Science. Federal University of Viçosa. Viçosa. 1996.

MORO, N.; AURAS, A. P. **Introduction to maintenance management.** Federal technological education centre of Santa Catarina metal mechanics educational management industrial mechanics technical course, www.norbertocefetsc. Florianópolis. 2007.

MURREL, K. F. H. **Ergonomics-man in this working environment.** London, CHAPMAN and HALL, 1975.

NEPSTAD, D. C. et al. **Avança Brasil:** the environmental costs for the Amazon. Report. Future scenarios for the Amazon project. Environmental Research Institute,

Socio-Environmental Institute. April 2000. Available at: http://www.whrc.org/resources/published_literature/pdf/NepstadAvanca. Accessed on 22 December 2011.

NOCE, R. et al. Analysing risk and return in the forestry sector: wood products. **Revista Árvore,** v. 29, n. 1, p. 77-84, 2005.

NOCE, R. et al. Export concentration in the international sawn timber market. **Revista Árvore,** v.29, n.3, p.431-437. 2005.

O'LEARY, J. E. **Some factors affecting the feasibility of helicopter logging in the Pacific Northwest and Alaska.** OSU, Corvallis, 1962. 108p.

PETZHOLD, M. F. (2011). The macroergonomic and political vision of traffic safety. **Revista Ação Ergonómica,** 1(4). 2011.

PEREIRA, D. et al. **Forest Facts from the Amazon 2010.** Imazon. Institute for Man and the Environment in Amazonia. Belém. 2010.

PEREIRA, M. P. **Fleet of vehicles used in forestry road transport in the municipality of Alta Floresta - MT.** Alta Floresta. UNEMAT. 2005.

PORTO, M. F. S. **Analysing risks in the workplace:** knowing in order to transform. São Paulo. Kingraf, 2000. 42p.

QUADROS, D. S. de. **Forestry Transport Workbook.** Regional University of Blumenau. Centre for Technological Sciences. Forestry Engineering Course. Roads and Forestry Transport. July 2004. P. 4, 26.

RAMOS, M. S.; VACCARO, S.; FERNANDES, R. S. The use of environmental perception as a tool for assessing the environmental citizenship profile of university students - National Conference on Environmental Legislation - CONLA. Univix. Vitória- ES. 2005.

RIVERA, E. F. R. **Road transport of logs in the province of Darién- Panama.** 2012. Available at: http://scholar.google.com.br/scholar. Accessed on 15 January 2013.

SANTOS, R. da S. **Level of health and quality of life of road transport drivers participating in the Uruguaiana-RS dry port.** Pontifical Catholic University of Rio Grande do Sul. Uruguaiana Campus Faculty of Philosophy, Sciences and Letters

Physical Education Course Uruguaiana-RS. 2008. P. 32.

SBS- BRAZILIAN FORESTRY SOCIETY. 2004. **Statistics: Brazilian forestry sector; socio-economic data.** Available at: http://www.sbs.org.br/setor_florestal.htm. Accessed on 28 November 2012.

SEIXAS, F. **Logging.** In: MACHADO, C. C. (Org.), Forest harvesting. Viçosa: UFV, 2002. p. 89-128.

SEIXAS, F. **New Technologies in the Road Transport of Wood.** Department of Forestry Sciences, ESALQ/USP. Piracicaba-SP. 1992.

SEIXAS, F. **New technologies in road transport of timber.** Brazilian Symposium on Forest Harvesting and Transport, v. 5, p. 1-27, 2001. SEO, N. Y. The operational logistics of road timber transport.
Opinions magazine. 2010. Available at: http://www.agrocim.com.br/noticia/A-logistica-operacional-do-transporte-rodoviario-de-madeira.html. Accessed on: 18 June 2011.
SILVA, E. P. **Evaluation of ergonomic factors in forestry extraction operations in mountainous terrain in the Guanhães-MG region.**
Dissertation (Post-Graduation in Forest Science) - Federal University of Viçosa-MG. 2007.

SILVA, K. R. et al. Evaluation of the profile of workers and working conditions in joinery shops in the municipality of Viçosa-MG. **Revista Árvore,** v.26, n.6, p.769-775, 2002.

SILVA, P. B. **Own or outsourced fleet:** which is the best option for delivering goods? Heavy Vehicle Maintenance and Logistics Supervisor Translog- Transportes e Logística - AmBev. Graduated in Financial Management. Uniuol-Faculdades. 2008.

SILVA, M. L. da. et. al. **Revista** da **madeira** - ISSUE N°117 - NOVEMBER 2008. Available at: http://www.remade.com.br/br/revistadamadeira_materia.php?num=1334&subject=Man. Accessed on 08 June 2011.

SIMÕES, J. W. et al. Formation, management and exploitation of forests with fast-growing species. Brasília, IBDF. 1981.139 p.

STEIN, F. R.; RODRIGUES, L. A.; SCHETTINO, S. **Road transport system at Celulose Nipo Brasileira - CENIBRA.** In: BRAZILIAN SYMPOSIUM ON FORESTRY HARVESTING AND TRANSPORT, 5, 2001, Porto Seguro. Proceedings... Viçosa, MG: Forestry Research Society, 2001. p.109-121.

TABOADA, C. M. Logística: o diferencial da empresa competitiva. **FAE Business Magazine,** n.2, p. 4-8, 2002.

TEIXEIRA, L. A. et. al. **Epidemiological analysis and correlation between capillary glycaemia and BMI in** truck drivers. Anais. XII Latin American Scientific Initiation Meeting and VIII Latin American Postgraduate Meeting - Vale do Paraíba University. 2008.
LAND. **Up to R$ 650,000: see the biggest lorries in the country.**
2012. Available at: http://economia.terra.com.br/carros-motos/. Accessed on: 15 December 2012.

TRINDADE, T. P. et al. Study of the durability of soil-rubi grade 81 mixtures with a view to application in forest roads and conventional pavement layers. **Revista Árvore,** v.29, n.4, p.592-600, 2005.

VALERIANO, S. **Primer on forest labour.** INTERNATIONAL LABOUR ORGANISATION - ILO AND THE BRAZILIAN FOREST SERVICE - SFB. 1ª Ed. 2009.

VELLOSO, F. A. M.; LOPES, E. T.; ROLDI, L. M. **"TRITREM" Alternative for transporting wood.** In: BRAZILIAN SYMPOSIUM ON HARVESTING AND FORESTRY TRANSPORT. 1997, Vitória. Proceedings. Vitória: Forestry Research Society, 1997. p. 157-175.

VELTEN, R. Z. et al. Mechanical characterisation of ground granulated soil-high slag mixtures for forest road applications. **Revista Árvore,** v.30, n.2, p.235-240, 2006.

VERÍSSIMO et al. **Logging Expansion in Amazonia:** Impacts and Perspectives for Sustainable Development in Pará - Belém. Imazon. 2002.

WINTERS, K. **Heart of MT suffers from grain transport on dirt roads.** 2013. Agroindustrial Logistics Research and Extension Group of the Luiz de Queiroz College of Agriculture (Esalq-LOG), University of São Paulo (USP). Available at: http://www.dci.com.br/agronegocios. Accessed on 05 February 2013.

ZAGONEL, R. **Analysing the optimum density of forestry roads on flat terrain in areas producing *Pinus taeda*.** Dissertation.
Master in Forestry Sciences. UFPR. Curitiba, p.100. 2005.

ZUANAZZI, D. M. **Survey and analysis of vehicular compositions used in forestry road transport in the Sinop-MT region.** Final course work. Institute of Agricultural and Environmental Sciences.
Federal University of Mato Grosso. Sinop. 2011.

ANNEXES

ANNEX I - Images of the vehicles used to transport logs by road in the municipality of Sinop-MT.

Trucks.

6-axle bimini truck (tractor+trailer).

7-axle bimini truck (tractor+trailer).

4-axle trailer (mechanical horse + semi-trailer).

5-axle trailer (mechanical horse + semi-trailer).

6-axle trailer (mechanical horse + semi-trailer).

Rodotrem 9 axles (mechanical horse+semi-trailer+trailer).

10-axle skid steer (mechanical horse+semi-trailer+trailer).

Images of the vehicles and their respective years of manufacture:

1980 1987

2006 2012

Safety equipment for transporting logs.

Vehicle without rear protection panel.

Figure 38 - Front protection panel missing.

Non-existent or non-standard pins and plugs.

Poorly distributed and excessive load.

Regular vehicles in terms of safety.

ANNEX II - Qualitative and quantitative questionnaire.

Ficha de Identificação do Veículo de Carga Transporte Florestal

1.0 Identificação do Veículo	Número de Identificação:	Placa:
Marca:_______________	Modelo:_______________	
Km Rodado:_______________	Ano de Fabricação:_______________	
Atividade Principal ( ) Florestal ( ) Agrícola	( ) Outra - Especificar:_______________	
% Horas Trabalhadas na Atividade Florestal: ( )100% ( )75% ( )50% ( )25% ( ) Eventual		
1.1 Motor e Capacidade do Veículo		
Potência do Veículo: _______________	( ) cv	( ) hp ( ) kWatts
Classes de Potência	Toneladas Máximas Permitidas	Potência Compatível
( ) leve	< 10 ton	até 60 hp
( ) médio	10 > x < 20 ton	até 120 hp
( ) semi-pesado	20 > x < 30 ton	até 180 hp
( ) pesada	30 > x < 40 ton	até 240 hp
( ) extra pesada	> 40 ton	maior que 240 hp

Visão do Operador:

Capacidade de Carga Em toneladas: _______________ Em m³: _______________

Direito de Propriedade do Veículo:

( ) Próprio - ( ) Quitado ( ) Não - Quitado ( ) Empresa Terceira

( ) Empresa ( ) Outra - Especificar:_______________

1.4 Manutenção do Veículo

Manutenção do veículo: ( ) Preventiva ( ) Corretiva. Faz Manutenção Regular? ( ) Sim ( ) Não

No geral, qual estado de conservação do veículo: ( ) Ótimo ( ) Bom ( ) Ruim ( ) Péssimo

Observações Adicionais

3.1 Atividade Presente:

Origem da Carga: _______________(Empresa)

Município de Origem da Carga:_______________ UF:__________

Destino da Carga: _______________(Empresa)

Município de Destino da Carga:_______________ UF:__________

Qual a Distância Percorrida __________km Tempo:

Ficha Pessoal do Condutor de Veículo de Carga Transporte Florestal

4.0 Dados Físicos/ Pessoais do Condutor

Altura/Estatura	__________m
Peso	__________kg
Envergadura	__________m
Fórmula:	
IMC = Peso (kg)/ Altura (m)^2	
Categorias IMC	
( ) Obes. Grau III	≥ 40,0
( ) Obes. Grau II	35,0 - 39,9
( ) Obes. Grau I	30,0 - 34,9
( ) Sobrepeso	25,0 - 29,9
( )Peso Normal	18,5 - 24,9
( ) Sub-nutrido	≤ 18,5

Envergadura ______ m

Altura ______ m

Peso ______ kg

Grau de escolaridade:

( ) Fundamental ( ) Médio ( )Superior

Estado Civil:

( ) Solteiro ( ) Casado ( ) Viúvo ( ) Separado/Desquitado ( ) União Estavél

4.1 Percepção sobre Permanência no Trabalho

Histórico de Profissões:
( ) Condutor de Veículos de Transporte de Cargas
( ) Condutor de Veículos de Transporte de Carga Florestal
( ) Outro -Especificar:____________________

Tempo de Trabalho:
Quantidade de anos que trabalha na profissão de transporte de cargas: ______________anos
Quantidade de anos que trabalha na profissão de transporte de cargas florestais : ________anos

Capacitação:
( ) Curso de Transporte de Madeira em Toras ( ) Curso de Direção Defensiva
( ) Curso de Segurança e Ergonomia na Operação ( ) Outro- Especificar:____________

Usa regularmente o cinto de segurança?
( ) Sim, utiliza com frequência ( ) Não,não utiliza com frequência ( ) As Vezes (*p.e.* Barr. Policias)

Você gosta trabalho que faz ? ( ) Sim ( ) Não
Permanência no Trabalho: ( ) Por comodidade ( ) Por gosto/opção ( ) Por falta de opção
Você mudaria de profissão? ( ) Sim ( ) Não, e Porque?____________________

Trabalho Cotidiano
Quantas horas trabalha por dia? __________hrs Quantas horas dirigi direto? __________hrs
Dirige durante a noite?( )Sim ( )Não ( )As vezes

4.3 Acidentes de Trânsito
Já teve problema com carga ( ) Sim ( ) Não Qual?____________________
Já se envolveu em algum acidente? ( ) Sim ( ) Não. Mortes ou Feridos Graves ( ) Sim ( ) Não
Considera sua carga perigosa? ( ) Sim ( ) Não Porque?____________________

4.4 Percepção do Veículo
O acesso à cabine é fácil relacionado a sua estrutura corporal (altura entre degraus, porta)?
( ) Sim ( ) Não ( ) OBS:____________________
A posição no trabalho é confortável?
( ) Sim ( ) Não ( ) OBS:____________________
O assento condiz com sua estrutura corporal (altura,peso)?
( ) Sim ( ) Não ( ) OBS:____________________
A cabine é de tamanho confortável?
( ) Sim ( ) Não ( ) OBS:____________________
O assento é regulável? Tem ajuste adequado para as pernas a altura?
( ) Sim ( ) Não ( ) OBS:____________________
O desenho e ângulo do assento e descanso para os braços são corretos?
( ) Sim ( ) Não ( ) OBS:____________________
O assento possui amortecimento e regulagens de altura e comprimento?
( ) Sim ( ) Não ( ) OBS:____________________
O nível de vibração/trepidação lhe incomoda? ( ) Sim ( ) Não ( ) OBS:__________
O nível de ruídos lhe incomoda? ( ) Sim ( ) Não ( ) OBS:__________
Existe sistema de ar condicionado na cabine?
( ) Sim ( ) Não ( ) OBS:____________________

3.6 Problemas de Transporte de Carga Florestal
Qual o maior problema do transporte florestal rodoviário de cargas? **(Problema Único)**

Ficha de Percepção do Condutor de Carga Transporte Florestal em Relação as Estradas

2.1 Percepção Sobre Estradas
Condições das Estradas de Terra
Estado de conservação: ( ) Bom Estado ( ) Médio Estado ( ) Ruim Estado ( ) Péssimo Estado
Condições das Estradas de Asfalto
Estado de conservação: ( ) Bom Estado ( ) Médio Estado ( ) Ruim Estado ( ) Péssimo Estado
Características: ( )Mão Simples; ( )Mão Dupla; ( )Possuem acostamento; ( ) Pontos de Apoio Bons
Velocidade Média em Asfalto: __________km/h Velocidade Média em Terra: __________km/h
Problemas das Estradas: __________, __________, __________, __________

2.2 Percepção Sobre A Frota
Avaliação sobre a Frota de Transporte de Veículos de Madeira em Toras (Estadual)
Estado de conservação: ( ) Bom Estado ( ) Médio Estado ( ) Ruim Estado ( ) Péssimo Estado
Avaliação sobre a Frota de Transporte de Veículos de Madeira em Toras (Nacional)
Estado de conservação: ( ) Bom Estado ( ) Médio Estado ( ) Ruim Estado ( ) Péssimo Estado
Avaliação sobre a Frota de Transporte de Veículos de Cargas
Estado de conservação: ( ) Bom Estado ( ) Médio Estado ( ) Ruim Estado ( ) Péssimo Estado

Printed by Books on Demand GmbH, Norderstedt / Germany